Numerology and Astrology for Beginners

A Soul's Journey through the Magical World of Numbers, Zodiac Signs, Horoscopes and Self-discovery

Crystal Hathaway

2

© **Copyright 2019 - All rights reserved.**

The content contained within this book may not be reproduced, duplicated or transmitted without direct written permission from the author or the publisher.

Under no circumstances will any blame or legal responsibility be held against the publisher, or author, for any damages, reparation, or monetary loss due to the information contained within this book, either directly or indirectly.

Legal Notice:

This book is copyright protected. It is only for personal use. You cannot amend, distribute, sell, use, quote or paraphrase any part, or the content within this book, without the consent of the author or publisher.

Disclaimer Notice:

Please note the information contained within this document is for educational and entertainment purposes only. All effort has been executed to present accurate, up to date, reliable, complete information. No warranties of any kind are declared or implied. Readers acknowledge that the author is not engaging in the rendering of legal, financial, medical or professional advice. The content within this book has been derived from various sources. Please consult a licensed professional before attempting any techniques outlined in this book.

By reading this document, the reader agrees that under no

circumstances is the author responsible for any losses, direct or indirect, that are incurred as a result of the use of information contained within this document, including, but not limited to, errors, omissions, or inaccuracies.

Table of Contents

Introduction **8**

Chapter 1: Getting Started with Numerology **10**

Changing Your Life **10**

The 3 Facets of Numerology.................................... **13**

Culture and Numerology.. **14**

The Significance of Your Name................................. **18**

Fibonacci Numbers **19**

Chapter 2: The Power of Numbers.......................................**23**

Decoding Numbers 0 to 9......................................**25**

Decoding 1111 ....**35**

Decoding 911 .... **36**

Decoding 1234 ....**37**

Decoding 1010 .... **38**

Decoding Numbers 111 to 555 .... **41**

Chapter 3: Numerology in Your Life **50**

Numerology for Your Health.................................. **50**

Numerology in Pregnancy **55**

Numerology for Positive Energy **58**

Numerology for Success **60**

Numerology for Relationships **64**

Chapter 4: The Beginnings of Astrology.............................. **69**

The Big Picture **69**

Astrology Through the Ages...70

Astrology and Your Life 72

Astrology and Psychology...73

Chapter 5: Astrology and the Planets 75

The Sun: Identity and Life Purpose 76

The Moon: Protection...77

Mercury: Learning 79

Venus: Love 81

Mars: Action 83

Jupiter: Growth 85

Saturn: Maturity... 86

Uranus: Change 88

Neptune: Imagination... 90

Pluto: Power 92

Chiron: Healing 94

Chapter 6: Understanding the Zodiac........................ 96

Interpretation of Each Sign..97

Correspondence of Each Sign 98

Compatibility of Each Sign 99

Aries 101

Taurus...105

Gemini..109

Cancer 113

Leo...116

Virgo 120

Libra 124

Scorpio 127

Sagittarius 131

Capricorn 135

Aquarius 139

Pisces 142

Conclusion *146*

Introduction

Throughout history, astrology and numerology have been sources of self-knowledge, wisdom, and connection.

Astrology was first developed in ancient cultures where humans viewed themselves as part of the entire living universe. They were not detached observers, but participants or players in the stage play of the cosmos.

Even in the year 3,000 BC in ancient Mesopotamia, people looked to the stars to help them gain special insights and understanding of the world around them. Back then, no one had a clear idea that the universe was infinite and forever expanding. Yet they looked to the stars and saw the infinite knowledge present within the stars' alignment and presence.

Over the millennia, astrology and numerology have become part of the various schools of thought that have permeated into various sections of both Western and other societies of the world. People in various locations of the Earth have tuned their minds to be receptive to the knowledge that astrology and numerology provide them.

Astrology and numerology are both products of an understanding that recognizes us as entities with a universe in which everything has meaning and everything is interconnected.

Presently, the main thought that is accepted by many is that we are isolated beings within the design of the universe and that life is a biological incident. There exists no inherent purpose or meaning. There is the date of your birth, a series of incidents that take place throughout your life, and finally, your departure from this world. One might think that the entire train of thought could be rather nihilistic.

This is where astrology and numerology come into play. They seek to understand our place in the wheel of time and our purpose in the grand scheme of things. After all, when there is meaning in our lives and with our actions, we yearn to live more. We want to explore this world further. We hunger for knowledge about the cosmos.

Purpose gives us drive. It gives us hope, and it allows us to dream. The purpose of our lives is not easy to spot. It is always coded and hidden.

Which is why these purposes need a tool that can help decode their mysteries. It's like the book "The DaVinci Code." It does not matter how many codes, paintings with messages, and secret vaults are present in the story. One always needs the knowledge to crack them. Fortunately, the protagonist is a professor of history and symbology. To him, the past and symbols are keys to solving all the mysteries around him.

We are the protagonists of our story. The symbols, messages, and cryptic codes are present all around us. To us, the knowledge of astrology and numerology are the keys to solving the mysteries of life itself.

Our story is about to start. The clues are laid out.

Let's understand them. Let's figure out the mysteries. Let's unlock its secrets.

Time to open our eyes.

Chapter 1: Getting Started with Numerology

Changing Your Life

Numerology is a method of gaining insight, knowledge, wisdom, and information about yourself and the world by using symbols in the form of numbers.

One of the things that you need to realize is that numbers were used well before words even entered modern language. Numbers also unlocked numerous mysteries of the universe. With some of the subjects that focus on the universe and its biggest unsolved problems, math is often involved to solve the puzzles. Physics uses math to make sure that its theories can be measured. Astronomy is nothing but a combination of physics, chemistry, and math all focused on the stars.

Numbers solve problems.

And this holds true in our everyday life as well. To numerologists, numbers present themselves as a language from nature itself. This language allows us to understand the psychological, philosophical, and scientific information littered all around us. We can even use numbers to understand the vibrations or energies that constantly adapt and change in the world around us.

It is for this reason that you can use the numbers that you discover in your life to guide your knowledge about your work, your love, your relationships, and other matters of your life. You can also use numbers to see the bigger things in the world around you. When you apply your inner wisdom and your logical reasoning, you begin to enhance your intuition.

Numerology can help you understand what you can gain from the journey of life and further inspire you to seek out its hidden meanings.

After all, life itself is made of a series of interconnecting circumstances and situations. Everything is about cause and effect. One thing causes a result, which creates an effect on something else, which in turn forms another cause, and so on. These causes and effects are linked to the many facets of this world, be they human, minerals, nature, animals, or the elements themselves.

Why?

Because any of the above facets can create a cause in your life.

Think of the butterfly effect, also called the chaos theory.

The question behind the popular theory is: Can the flap of a butterfly's wings in Mexico cause a giant storm in Texas?

Physically, that is highly improbable. The chances of the above situation occurring are minuscule. But the situation is an example of stretching our thought processes and understanding how a small action in our lives can lead to big consequences.

Each point in your life has multiple branches. Every action that you take creates a cause. And this cause could be one of many causes that you could have created. Should you wake up earlier for work or should you simply grab another half-hour of sleep? When you make your choice, you create an effect. Perhaps, if you take some extra time to catch some z's, you might end up missing the bus to work, which in turn could cause you to arrive late. If you hurry, you might forget that it is your dad's birthday today. You were supposed to wish him a happy birthday first thing in the morning. Whoops! Now you call your dad, but he is not available. You decide to call him after work. But eventually, after a tiring day's work where you are mostly playing catchup to your tasks, you head home, feeling exhausted.

Eventually, you are about to head to bed and you realize that you haven't wished your dad happy birthday yet! You hurry to the phone and call him. He is, of course, happy to hear your voice. He loves you, after all. But deep down, you know that you paid more attention to your work than your dad.

From that day onward, work takes on a different meaning to you.

That is what the butterfly effect is all about. The flap of the butterfly's wings was you waking up late. The hurricane was the resulting effect, your eventual distaste for your job.

Now try and think of the many little flaps in your life and how they changed your life to bring you to where you are right now. Try and think back and imagine what would have happened if you had made one little change in your life? What if you had actually gone to that job interview? What if you had said yes to that proposal? What if you had chosen to stay closer to your parents?

What if?

There are many such situations scattered throughout your life. Yet all of them happen to you without revealing what the bigger picture is.

That's where numerology comes into play. It peels back the curtains to give you a glimpse of what is really happening in the background of your life. You might not get all the details, but you get a picture with more information than you have right now.

And with that information, you can change your life.

The 3 Facets of Numerology

When we look at numerology, we see that it has three unique facets.

The first facet is numerology itself, which involves the discovery and understanding of numbers as symbols for patterns or vibrations deeply seeded in the nature of things.

The second was discovered by the ancient Greeks–Arithmancy. Arithmancy is the set of methods used to work with the numbers and remove the meaning from their various connections. Arithmancy gives us the power to find the meaning we need from the chaos of numbers around us.

The third and final facet of numerology is what the Greeks call Isopsephy. The Hebrews had another name for it: Gematria. But both names and ideas represented the same aspect of assigning numbers to the alphabet or, in some cases, the phonemes of a language. These methods help numerologists transform words and names to numeric characters.

The three facets act as a tripod for numerology. They cannot function on their own. If you remove one leg, it tips over the nature of numerology and you are left with half-understood meanings and undeciphered messages.

But when you combine the three facets, then you are able to understand not just your life, but your strengths. You are able to face your weaknesses with more clarity. You may find challenges that have solutions.

Culture and Numerology

Numerology is part of almost every culture. However, certain cultures around the world such as the Tibetan, Hindu, Chinese, and Greek, have taken numerology a step further. They have developed techniques and methods that dig deep into the core of knowledge.

Each culture has its own unique approach based on its history, collective energies, and the level of consciousness of the people following that culture. For example, in certain parts of the world, people spend their entire lives within a certain region or area. To them, the interpretation of numbers might be different from those who live in urban areas. In other words, the meanings of the numbers have a much different effect on them than people living in cities. The idea of understanding numbers that repeat themselves every day might be different for a person in a metropolis who is constantly checking the time to be in tune with the various schedules of his or her day.

On the other hand, someone who lives in the countryside, a small town, or even a village might face a different scenario, especially if he or she is not concerned with time down to the minute, but rather time as viewed from the perspective of a day. In other words, he or she might think of the day as the morning, afternoon, and evening and what happens during those periods. A person living in the city might view time as 9:00 am, 1:20 pm, and so one, placing various actions and decisions on more specific time slots.

However, despite the many ways numerology can be interpreted, there is still an overarching message behind the various numbers, patterns, and numeric symbols we see every day.

To numerologists, these messages are important. But they wouldn't have been able to gain the knowledge of numerology if it weren't for

Pythagoras. The Greek philosopher is often shrouded in mystery as little of his original work survives today and many people who wrote about him did so nearly a hundred years after his death. What is known is that he was deeply involved in philosophy, music, and, of course, mathematics.

In grade school, you might have come across the Pythagorean theorem, which states that in a right triangle, the length of the hypotenuse can be derived from the sum of the square of the remaining sides. Apart from this popular mathematical formula, Pythagoras was also deeply vested in the mystical characteristics of numbers.

He began to dive deeper into the mystical aspects of numbers after discovering something interesting: A series of consecutive odd numbers beginning with 1 always gives you a number that is a square of a number.

Let's try it out.

$1 + 3 = 4$ (which has a square of 2)

$1 + 3 + 5 = 9$ (which has a square of 3)

$1 + 3 + 5 + 7 + 9 + 11 + 13 = 49$ (which has a square of 7)

You can continue adding consecutive numbers to the series and you will always end up with a number is a square of another number.

Isn't that fascinating?

Pythagoras thought so too. And it was this concept with numbers that sent him on a journey to discover more meanings behind numbers. Eventually, it led him to coin the phrase 'all is numbers.'

There are many interpretations of that phrase, but one theory suggests that it refers to the idea of describing everything in the world in terms of numbers.

If you think about it, that is not so far from the truth. After all, science and mathematics both depend heavily on numbers. Think about it. Engineering depends on numbers. Biology depends on numbers, as well (heart rate, blood pressure, white blood cell count, and more). Pick out other areas of science and you will notice that numbers play a vital role in their functioning.

Pythagoras continued to examine numbers through non-numerical methods. This meant that he was looking at numbers through the lenses of intuition and mysticism. Here are some of his findings.

During his studies, he suggested that the number 1 was creative. This is because you can keep adding ones to create any number. Of course, you might need a lot of ones to create a number like 1,000,000 (one million ones to be precise), but the theory still stands. You cannot create 3 by just using the number 2. You can create 9 by only using the number 5. But 1 is dynamic. It is creative.

He recognized the number 2 as female and the number 3 as male. When combined, 5 denoted marriage. Since 5 appeared in the middle of the sequence of numbers from 1 to 9, it also represented justice.

The number 10 was considered a holy number. Because of the presence of the creative number 1 and a null number 0, 10 represented the opposites of things. In other words, it became the yin and yang of nature. Some of the opposites it represented were:

- One and many

- Straight and crooked

- Odd and even

- Good and evil

- Light and darkness

- Rest and motion

- Masculine and feminine

- Right and left

- Limited and unlimited

After the death of Pythagoras, the focus on the mystical properties of numbers began to wane. People paid attention to numerology less and less as the years passed.

It was not until the 1800s that a certain numerologist by the name of L. Dow Balliett wrote many books on numerology. While she was not the first one to focus her attention on publishing books on the subject, she was definitely one of the few who incorporated the teachings of Pythagoras heavily. According to her and other modern numerologists, each number has its own specific vibration. Along with numbers, even objects, people, food, sounds, and colors also have their own vibration. **If someone wishes to live in harmony with their surroundings, then he or she must match the rhythms of his or her vibrations to those of the environment.**

As time went on, more and more studies were conducted on numerology. It was connected to the names of people, where each name was given a numerical representation. This is because numerologists considered the name given to a person during birth to be significant. It is much more significant than the person's name after marriage (if the person chooses to change names then), nicknames or any other name picked up during the course of life.

Numerology also began to be associated with understanding various aspects of life including areas such as the career path that fits the person, positive attributes that the individual can focus on, negative traits to modify or eliminate, and the ideal traits of a romantic partner.

In short, numerology became a way to unravel the confusion or answer the questions to many of life's complex problems.

Numerology also began to debunk many of the beliefs held about certain numbers. For example, the number 666 was often associated with the devil in many Christian traditions. However, if you took the concept of numerology and applied it to people's names, then all you have to do it tweak the names a little to make it seem like practically anyone's name can total 666.

As numerology began to develop further, it began to undergo a system of refinement, where more theories and studies were conducted to improve its application.

These changes eventually bring us to the numerology that we know and understand today.

The Significance of Your Name

We skimmed through the idea that our name holds significance in numerology. Why is that? What makes our name so special?

If you enter the name David Jones into a Google search bar, you are going to receive plenty of results for it. The name is rather popular, and there are a number of David Jones around the world. Similarly, the name Jane Fonda provides you with a number of search results, not limited to the popular actress and political activist. In short, we could share our name with hundreds–maybe even thousands–of people around the world. So why are we special?

But what about Pilot Inspektor?

Oh, you think that is not a name? You will be surprised to know that it is. In fact, what if I told you that the name belongs to the son of a once-popular American actor? Indeed. Ever heard of Jason Lee? Well, his son bears this rather unorthodox name (which was apparently inspired by the lyrics of a song). Now, I am not judging the choice of name, merely drawing a very important conclusion.

Whether your name is unique or common, it is still unique to *you*. Whether you are David Jones or Pilot Inspektor, your name is still yours.

You and your name are just one combination that does not exist anywhere else. It's like a fingerprint; no two sets are identical. In the same way, you create an imprint on this world that cannot be made by any other person. Even if you had a twin who shares the exact birth date and time, has the same physical features as you, and even bears your name, you are still unique.

Your choices and your actions decide where you go in life. Each step you make is unique to you. Which is why numerology considers your name to be significant.

Fibonacci Numbers

Before we delve into the significance of the Fibonacci numbers in numerology, let us try and understand just what the numbers represent.

In mathematics, Fibonacci Numbers follow the below sequence:

0, 1, 1, 2, 3, 5, 8, 13, 21, 34, 55, 89, 144, and so on.

Each number is the sum of the previous two.

1=1+0, 2=1+1, 3=1+2, 5=2+3 and so on.

Now that sounds like just about any cool mathematical sequence, doesn't it? After all, anyone could have found that out. So what exactly makes the sequence of numbers unique? What is considered special?

This is a sequence that appears in nature. The Pyramids of Giza have the Fibonacci sequence. Insects and flowers have this sequence as well. Eventually, the sequence of numbers leads to a ratio 'The Golden Ratio.'

This ratio explains symmetry in nature.

However, scientists are still baffled as to why these patterns exist in our world. It is true that not everything in the world follows the Fibonacci sequence. There are various numerical patterns to objects and entities that surround us. But what makes this particular sequence significant?

One of the theories that scientists are pursuing is that the golden ratio exists in nature because that is the most effective growth pattern that nature prefers. For example, those plants growing with the Fibonacci sequence in their design allow for maximum arrangement of seeds and exposure to sunlight. In human beings, scientists who study evolutionary advantages have noticed that the ratio is used to define physical attraction, as the more a face incorporates the symmetry of the Golden Ratio, the more appealing it looks.

For this reason, the sequence plays an important role in numerology to discover the most ideal path that we should take in our life. It guides us toward choices and actions that are not only suitable for us, but help us develop more in our lives. It provides us the opportunity to take maximum advantage of the good fortune presented to us and to attract even more when the time comes.

That is why numerologists combine the power of Fibonacci and your birth name. The idea behind this is twofold:

- The number sequence provides an insight into your life so that you may be able to take the most ideal path toward a particular outcome. But how can you be sure that the insight or guidance is meant for you? What if it could be for an unknown John Doe living halfway across the world?

- This is where your name comes into play. By focusing on your name, numerologists ensure that the guidance that you receive is meant for you only. When they bring together the letters of your name and correlate it to the Fibonacci sequence, they are creating a connection between your name and the sequence.

With that, you could even say that just like your name, the Fibonacci sequence for your name is unique as well.

Using the Fibonacci Sequence

An important requirement for using the Fibonacci sequence is the arrangement of letters.

1	2	3	4	5	6	7	8	9
A	B	C	D	E	F	G	H	I
J	K	L	M	N	O	P	Q	R
S	T	U	V	W	X	Y	Z	

Using the table above, you can determine the Fibonacci number for your name.

Let us take an example of a person named Michael Lowrey.

The name will be split to represent the following:

M+I+C+H+A+E+L+L+O+W+R+E+Y

Using the table above, we can substitute the letters with the numbers.

$4+9+3+8+1+5+3+3+6+5+9+5+7 = 68$

$6+8 = 14$

$1+4 = 5$

Your Fibonacci number is 5.

Chapter 2: The Power of Numbers

Noticing numbers all around you is not special. After all, countless messages are projected through numerical representations. Telephone numbers, area codes, discount percentages, prices, and plenty of other information are number based.

But what if you start noticing a pattern to these numbers?

Think of this scenario.

You wake up in the morning and notice the time on your clock. It's 10:10. You are late for an appointment!

You get dressed quickly and head out the door, a piece of doughnut in your mouth the only sustenance you are going to get for that morning. You are holding your phone in your hands, simultaneously trying to order an Uber to get you to your destination. As you punch in the address, you notice that location says 10th Lane on 10th street. That's rather odd. But you shrug off your thought processes and chalk up the number sequence to coincidence.

You exit your building and are waiting for the ride. Eventually, you spot the car turning the corner onto your street. Your eyes are drawn to its license plate, and you see 1010 on it.

Okay, now that is definitely more strange than usual.

Before you can make any connections, the driver pulls up in front of you and you get into the car. After the initial pleasantries, you settle back into the seat and focus your mind on the weird repetition of numbers you have been experiencing. Eventually, you notice the driver takes out his wallet and opens it to get something. You are given a peek at his driver's license and you notice the numbers.

1010 is among them.

What is happening? Do you need to pinch yourself so you can actually wake up? Is this still a dream that feels so real?

Let us get one thing out of the way. What you are experiencing is not a dream and is no mere coincidence. Of course, to most people, it would indeed seem to be a random repetition of numbers.

But in the world of numerology, these numbers are a message. They mean that this universe is placing something in your path.

Let us take a step back here because one question that people often ask is why doesn't the universe simply come out and give us the knowledge that we seek? Why can't it just say, "Hey! Don't go there because you might waste your time at that job interview and miss out on the other opportunity that you had received. Also, the interviewer has a powerful cologne that will send blistering headaches searing through your brain. Might I also wonder, did you forget some loose change today? You might need it for the toll so stop at the coffee shop on 11th street to get yourself a decaf and some change. No caffeine. Doesn't go well with that vodka you had yesterday."

Wouldn't that be so convenient? Well, there are two reasons why you don't get messages in such an obvious manner.

- The knowledge of the universe is powerful and we are not capable of handling it. It is similar to the end sequence in Indiana Jones 5 (spoiler alert) where the villain asks for more knowledge. Eventually, she receives so much knowledge that it literally burns her body from the inside out. Not a pleasant feeling, I must assume.

- The second reason is that the universe does not want to directly influence our decisions. If it does so, then it is removing the act of free-will from our lives. Imagine knowing that everything you do is controlled by the universe. It's like we are puppets attached to strings and the universe is controlling us in this giant play taking place all over the Earth. This is why, rather than directly sending messages, the universe gives you code. That way, you get choices. You can choose to decipher the message or ignore it. Once you have deciphered the message, you can choose to take it into consideration or ignore it once again. It's like the two-step verification process when you use your credit card online. Essentially, it is like the system (in our case, the system is the universe) asking: Are you sure you would like to go ahead with the action?

Now that you understand that, let us proceed further and see what each of the codes mean.

Decoding Numbers 0 to 9

1

Brief Meanings

This number represents the idea of independence, leadership, and originality. When you notice this number in your life, it means that either you are going to start something new or that a new event, idea, object, or person is going to affect your life.

The Detail of the Number

Pythagoras often referred to the number 1 as the 'limited and unlimited.' This is because one is a limit to itself. For example, one can be a number by itself, and it can stop there. It can also be unlimited due act that it can make other numbers. As we discussed before, two 1s give us the number 2, five 1s the number 5, and so on until infinity.

In numerology, one is considered as a leader. It is a vertical line and stands with a sense of purpose and confidence. This makes 1 the number of leaders, guides, and motivators. It shows the act of planning and taking charge.

As a person, you will aim to be the best. You are fiercely competitive and incorporate great physical energy. By physical energy, we are not talking about 6-pack abs! Rather, it means that you have the energy to take action. You do not like to lie around and wait for things to happen.

2

Brief Meanings

This number represents the idea of sensitivity and cooperation. This could mean that you are either a person who enjoys cooperating with others and is sensitive to their conditions or that you might need to be more cooperative and sensitive.

The Detail of the Number

Two can be considered the number of togetherness. It shows the act of caring, where it describes the person with the number 2 as one who goes out of his or her way to be there for someone else.

If you start noticing this number frequently, then you might be suffering the consequences of a troubled relationship, partnership, or the ability to cooperate with someone.

You might be alienating yourself from those close to you or from people in general. This means that you are probably figuring out how to resolve a situation in your life or you feel impacted by something and you need space to deal with it.

However, you are generally a loving and sensitive person. It could be because of this that you feel greatly affected by situations to a point where you feel compelled to simply work with them alone. The occurrence of the number shows that you have to break out of your shell. The fact that you are using your sensitive and loving nature to cause more harm to yourself is a waste of potential. Rather, you should be using it to show others your wonderful nature; one that brings more people to you. When you eventually attract people towards yourself, you are able to resolve many of your problems with ease. Cooperation helps bring more minds to solve the problem.

3

Brief Meanings

When you notice this number, it is usually a sign of self-expression. The number 3 also focuses on the aspects of creativity and the ability to produce unique results in works of arts. People with the number 3 are also seekers of spirituality.

The Detail of the Number

One of the unique aspects of this number is that it might not present itself directly. Rather, you might notice certain occurrences that show you that this number is present in your life.

The number 3 often appears in the form of actions or situations. You might notice that when you are headed outside, you pack exactly 3 important items that you think you might require. You have 3 cups of coffee every day. You might start putting 3 alarms to wake you up in the morning. When you are preparing coffee in the morning, there are 3 packets of sugar.

In similar ways, 3 is a mysterious number that makes you stop once you begin to notice its appearances.

This number is essentially telling you to pursue some creative endeavors. Your ability to express yourself has been greatly hindered and for that reason, you have been missing out on opportunities to do something big. In other words, the number is asking you to see if there is a possibility that you can choose to turn that creative ability of yours into a talent you can use.

4

Brief Meanings

With this number, you are looking at stability in your life, whether it is in your personal sphere or your professional sphere. It is also a number that represents a career and the opportunities tied with it. But on the flip side, it could also serve as a warning about your current professional situation.

The Detail of the Number

You might be working a nine-to-five job. If you are, then you might be feeling boxed in or claustrophobic. There is a sense of hopelessness where you feel that each day is like the other. You have no challenges, and there is a sense of boredom creeping into your life.

4 is the number that reminds you of that boredom and lack of accomplishment. When it appears to you, try and evaluate your professional life. Try and see if you have accomplished anything and if there are ways for you to do that. It could also be that you are finding it difficult to get along with people in your job. If you are, then it is time

to make some remedies.

Alternatively, it could also be that you are an entrepreneur or a business owner and you feel that there has been no progress. Sales have been stagnant or low. There is a lack of innovation in your work. You feel like your product or service is going toward inevitable doom. In such situations, 4 is trying to let you know that perhaps the best plan at that moment is to try and examine your product or service from a different perspective. Maybe you should start looking at the feedback you have been receiving and see if you can make changes to accommodate that feedback. You cannot obviously listen to each and every piece of feedback, but perhaps you could find out a general reaction among your customers.

One thing that you should remember is that if you have been planning for the future and have been completely engrossed in that planning process, then it is time to take a step back. You might need to work on what you have now before you think about what you should do for the future. If you miss out on the opportunities in the present, then you might not have the future that you desire.

5

Brief Meanings

Adventure is in the air. There is a need to break free from life and simply go out on a wild and exciting adventure. Maybe even travel the world. This also indicates a craving for excitement, where you simply want to do something different. However, this number is a sign that there are good things about to happen.

The Detail of the Number

Ever felt like you were working toward something, but there don't seem to be any results to show for it? The occurrence of the number 5 means that you shouldn't give up. At least, not yet. Continue on the path that you have set and give it all you have. The end might just be around the corner, and it would be disappointing if you give up when you were almost going to attain your goal.

Additionally, you might have been facing a particular challenge or a tough situation for a long time. So the presence of this number means that things are looking out for you. You might be about to get a reprieve that could be brief or long.

One thing to remember is that this number is not an indication that you have to quit your job. Some people assume that 5 is a sign that they are placing their hard work in the wrong avenue. While 5 definitely does not recommend that you keep pushing yourself in your job for long, it is definitely telling you to try for a little while longer so that you do not miss out on your rewards after putting in so much hard work already.

Finally, when you notice 5, you should try and do something that is exciting and adventurous during your reprieve period. Make the most out of your break. Do not simply spend it relaxing at home. Do something new. Travel. Experience something you would never have dared to before.

6

Brief Meanings

This is the number of family and the bonds that you share with them. It also represents the idea of a home. This may not be a physical structure, but it could also represent an area or a group of people who make you feel at home. 6 reminds you of your responsibilities and how you should never shy away from them.

The Detail of the Number

There are multiple reasons for the presence of 6 in your life:

- Perhaps you have been having trouble maintaining relationships with various family members. Or perhaps you have drifted away from each other. Whatever the reason, 6 is a reminder that it is time to get back and re-forge the ties you have with your family. Fixing things with your family is going to help you greatly in the near future.

- If you have been looking for a place to call home, then 6 is simply letting you know that home is not a physical structure. Perhaps you should see if you like the city or town you are living in. Maybe your friends are incredible people. In some ways, you might already be at home.

- Finally, it could be that you are feeling burdened by your responsibilities. But 6 is there to remind you that you are probably the only one holding things together. In other words, if you let go of your responsibilities, then those dependent on you might be affected greatly. At the same time, you might receive help to alleviate the pressure from you.

7

Brief Meanings

7 represents the path of the spirit. This number is a reminder for you to get in touch with your spiritual side. If you are already on a spiritual journey, then the number is asking you to renew that journey with vigor.

The Detail of the Number

Sometimes, we need to find a way to handle all the pressure, anxiety, stress, and other mental challenges that we face every day. Often, we become so lost in these mental challenges that we forget to look after ourselves.

7 is the wake-up call that you have been waiting for.

The first thing that it is trying to say is that you might need to clear your head. There is a lot of chaos in your mind that is preventing you from achieving your goals or giving the best of your abilities.

The second thing that 7 is reminding you of is the fact that you might be missing out on a lot of opportunities because of the fact that you have not been taking care of yourself. Your mental state is cloudy, preventing you from looking at blessings and opportunities. Your mind is pushing you toward dark thoughts.

Finally, you matter in this journey of life. The main focus of 7 is not on spirituality but on the spirit itself. And the spirit suffers when the mind and the body suffer. This is why 7 is the proverbial splash of water to the face. It is letting you know that if you don't take care of your mental and physical health, then you might be facing some serious problems in the future.

8

Brief Meanings

If you are looking for abundance, then 8 is a good luck charm for you. Apart from that, this number also indicates the presence of material prosperity or the possibility that such prosperity might be entering your life soon.

The Detail of the Number

We all love materialistic things to a certain degree. It could be that brand-new car or those stylish shoes that we spotted in the mall. Whatever it is, there are some things that we yearn to own.

8 is the indicator that you are not far from owning that object or material that will bring a sense of accomplishment or satisfaction in our lives. At the same time, 8 is also referring to the idea that if more is what you seek, then more might be what you are going to receive. But that means that you should not be slacking at this time. In other words, do not stop putting in the hard work for the things that you would like to receive.

If you have been experiencing the loss of things lately, then 8 is a reminder that such losses are going to end. The universe is going to reverse-engineer your life so you start attracting instead of losing.

9

Brief Meanings

This number is the universe's way of reaching out to your soul. Perhaps you have been living in a fog lately. Maybe you have gotten used to the things that are not working for you. Or maybe you saw an opportunity in life and you passed it, thinking that maybe you are not deserving of it. 9 might think otherwise.

The Detail of the Number

Wake up!

That is exactly what 9 is trying to tell you. You might be thinking that your life has no more wonderful surprises to give you or that your situation is not going to improve. When you start seeing 9, then you have to brace yourself.

This is because 9 is asking you to boldly take the step that you were afraid of taking. It is reminding you that there will be no changes in your life if you are not willing to take action to change anything. If you have accepted your life for what it is right now, then that is what it will be. However, if you are willing to do the real work to divert your life's journey on a different path, then you are more likely to be rewarded.

It does not matter if you have to take a new hobby or find a new passion in life. It does not matter if you wanted to ask that person out on a date or treat yourself to something nice. Time for you to snap out of your old thought processes and adopt a new skin, one that allows you to make changes in your life.

0

Brief Meanings

0 is the number of infinite possibilities. No matter what goals you want to achieve, there are so many ways you can do it, and you might find success in most of the ways you choose. One of the things that make 0 unique is that it has no end and no beginning. This means there is a void waiting for you to fill it with potential.

The Detail of the Number

You might notice 0 all the time. In fact, the number is fairly easy to spot. Just step outside and you might see zeroes everywhere. Many people think that they are actually seeing a sign from the universe when it is actually just another number.

So how do we know that the zero we are seeing is indeed a message from the universe and not a random appearance of the number?

We know this when the zero repeats itself.

In order to remove the veil of ambiguity, the universe will reveal this number to you as 000.

When you start noticing three zeroes repeatedly, then you know that you just got hit with an instant message from the universe. Make sure you were tuned in.

The main purpose of 0 is to let you know that it is time to explore your potential. It is letting you know that you have to stop limiting yourself. If you do not take chances, then you are not going to accomplish anything. Time to go out and grab all the opportunities that you have been presented with.

0 also reminds you that you are a unique individual who is part of a great connection. We are all part of the same giant web that this universe is. You are not alone. You are not weak. You are not powerless.

Decoding 1111

This number has the mystical property of connecting you to the universe. Many numerologists interpret this number as a type of 'Angel Number.' In other words, the number allows you to connect with the guardian angels that look out for you.

When you find repetitions of this number, then you first need to go back and look at the meaning of the number 1. We are talking about the ability to lead and take charge. We are also talking about the power to be confident and make challenging decisions. The number also talks about new beginnings.

The first message that we can take from the occurrence of 1111 is that there is a new change or beginning about to take place in our lives. But we have to be prepared to lead and make tough decisions with confidence when the time comes. We should not waver in our path and be certain of our actions as only through leadership, confidence, and certainty can we make the best use of the new opportunities that will come our way.

It is also possible that you might have missed a wonderful opportunity. The appearance of 1111 means that you have to start taking charge of your life. Evaluate your decisions and actions in the past few days or weeks. Try and see if you had received an opportunity and you had failed to take it when it had arrived to you. If you had, now would be the time to muster your confidence and return to that action or decision. And this time, be the leader. Take charge and make a new change into your life.

Additionally, this number represents the idea of low self-esteem. What this means is that you have been allowing other people to dictate and guide your life. This leaves you powerless, and you might have missed the opportunity to grab on to a new change. By noticing this number, it is time to remove negative thoughts from your mind.

You are the master of your journey. Only you can live it. Nobody else is going to live your life. So banish the low self-esteem and take charge.

Decoding 911

When you are in trouble, you dial 911. It does not matter where you are in this world. You have your own version of '911,' whether you have to dial 100 or 999. Essentially, the presence of 911 is a symbolic representation of calling for help.

Your life is in disarray and you have dialed the universal help hotline. Prepare yourself for help.

As we have seen, 9 refers to the accomplishment of your true purpose. In other words, it calls for you to act on what you really want to do right now. It also talks about giving passion and honest service in the profession that you have taken or the arts you have chosen.

Essentially, you are thinking about changing careers and you want to know if you are supposed to stay or if you are meant to leave. But the most important question that you have to ask yourself is: Are you doing what you are meant to be doing? If not, are you willing to put in your passion and honest service to get what you want? Are you dedicated to going the extra mile to be where you want to be? If the choice of change is risky, then are you willing to improve in the job that you are in now?

This can also be applied to other aspects of your life such as love, friendship, and habits.

The most important questions you have to ask are:

Are you happy where you are? If you want to change, are you willing to go the extra mile for it?

When you have answered the question, then 911 is essentially telling you to go ahead and do it. When you have truly found an honest answer, then you are meant to pursue it.

Decoding 1234

1234 is a number that follows a series. Each number is like a step toward a bigger number. In fact, the best way to interpret the number is by thinking of it as a series of stairs.

You start at the first step, climb to the second, proceed to the third, and finally land on the fourth.

In other words, 1234 is about progression.

Think of the appearance of this number as an answer to a situation that does not involve the need to look for 911.

In other words, you are comfortable where you are and happy with your position. However, you are doubting yourself. The seeds of doubt start small until eventually, they consume all your actions and lead to the subversion of your talents and abilities.

You start showing poor performance in your life.

1234 is the universe's way of telling you that you should stop with the self-doubt.

Is there a colleague at work who is suddenly getting more praise than you? It is not an indicator that you are poor at what you do. Stop doubting yourself.

Do you feel like your partner is suddenly happier than you? Perhaps something nice must have happened to them. Don't doubt your life.

Did someone criticize your talents? Don't immediately jump to conclusions. It could be an isolated incident or merely a sign for you to develop yourself to reach your full potential.

Remove the self-doubt from your life.

Decoding 1010

If you notice pictures of watches, you might often notice that their hands are in the 1010 position.

Why do watchmakers choose this position to display their watch and clock models?

- The hands do not overlap with each other. This allows the viewer to clearly see the watch or clock.

- The arrangement of the hands shows symmetry. As we saw earlier, symmetry is appealing to people and also falls as part of the Fibonacci sequence.

- It allows for the manufacturer to place their logo clearly on the watch or clock and even display any additional features of the timepiece more clearly.

The above concepts hold true for the occurrence of this number. But first, we have to decode each of the digits present in the number.

When you start noticing 1010, the number is holding the powers of both 1 and 0:

- 1 represents power, leadership, and confidence.

- 0 is the idea of nothingness. It breaks barriers and asks you to go past your limitations. It talks about using your full potential.

Using the above references, here is what the number 1010 means:

- Your path is clear before you. You have made a choice or are about to make another. Do not falter in your steps. If you feel that it is right, take charge and give your full potential into that choice. For example, let us say that you have chosen to move to another country. You should be confident about your choice and explore the opportunities it presents with all your capabilities.

- The idea of symmetry plays true here. Both 1 and 0 have to be expressed in harmony. You cannot be a leader and work half-heartedly. In the same way, you cannot be confident and avoid taking all the necessary actions toward your goal.

- When you are using both the powers of 1 and 0, you are allowing yourself to express additional characteristics and traits about yourself. Let us say that you are going for a job interview. You are confident and you are ready to show your full potential. This combination of 1 and 0 allows you to express more. Are you a talented orator? Do you have something else you can bring to the table? Do you have a diverse set of skills? Only through the expression of 1 and 0 are you able to put all of your additional capabilities on display.

Decoding Numbers 111 to 555

111

The number 111 does not imply a less potent version of 1111.

Rather, it is a combination of two numbers; 1 and 11.

As we have seen numerous times, 1 is the number of leadership. It signifies sureness and the decision to go after your actions.

But then we have the number 11. This number represents the idea that people do not like to work with the mundane. In other words, it talks about inspiration and a new way of looking at things. Some of the professions that can be taken as an example for this number are musicians, painters, writers, teachers, and actors. To them, the idea of doing the same thing over and over again does not fit well with their purpose in life.

In a similar way, 111 is about you taking the determination to look for inspiration in your life. It does not matter what situation you face, you need to try and, as they say, "spice things up."

Do you feel that you are having more friction in your marriage? Spice things up.

Do you feel that your life has been dull? Travel somewhere and spice things up.

Do you think that you are missing out on achieving something? Take up a hobby and yes indeed, spice things up!

But why is it necessary for a sign to tell you to look for inspiration?

Why can't you do it on your own? Isn't it obvious that when someone does not feel like they are doing anything of value, they should look for something exciting to add to their lives?

Here is the problem. People get comfortable with the situation that they are in. It is like that old adage that says that the devil you know is better than the angel you don't. When people become acclimated to a certain way of life, situation, person, or even idea, they do not look to make things interesting, even if they find themselves in a monotonous existence.

It takes the universe to jolt people out of their reverie and make them wonder: Why am I just letting things happen? I need to take charge of my life to make it interesting.

222

Just like 111, 222 is a combination of two numbers.

We have the number 2 which denotes cooperation and sensitivity. And then we have the number 22.

22 represents the idea of personality. It represents positive ideas such as a smile, kindness, modesty, gallantry, and compassion.

At some point in our lives, we all doubt the way we are. We often wonder why the people we are not have the traits that we are proud of carrying forward.

Think of it this way: If you were suddenly 12 years old, would your current self be something that your 12-year-old self can use as a role model? Would you be willing to inspire the younger version of yourself with the traits you have right now?

Let us look at how we can approach ourselves with the number 222.

The first part of it asks the question: Are you sensitive enough? When you do not have any sensitivity, then it means that you lack empathy. These days, you might have heard of a concept called 'emotional intelligence.' What the concept boils down to is that leaders need to have high emotional intelligence (commonly referred to as EQ) along with their IQ. The high level of EQ is what decides whether the leader will be able to successfully lead the people below him or her. One of the important emotional traits that the leader needs to have is empathy. Empathy shows how well the leader can understand others.

In the same way, with enough sensitivity, you will be able to have better empathy. When your empathy is high, you will be able to tune into the emotions and situations of others. This invariably allows you to cooperate with people better.

The second part of the number asks you to take stock of your emotions. Have you been feeling angrier lately? Have you been going through a period of depression? Has tragedy struck your life and are you going through an immense level of sadness?

All of the above emotions are negative emotions. They cripple the normal functioning of your life and affect you in many destructive ways.

222 is asking you to evaluate your emotional self. It is asking you to check the balance between your negative and positive emotions. Has the balance tipped toward one side? If so, which side and why?

Let's say that you have been experiencing depression lately. It has reached a point where you are suddenly plagued with dark thoughts. When you start noticing the number 222, it is a signal to look at your emotions and accept that there is an imbalance. In the case of depression, it is time to seek help and improve yourself. Seek out your friends or family. Visit a therapist. Look at making positive changes to your emotions and, eventually, your life.

333

This number is not a combination of two numbers. Rather, 333 should be taken as a whole.

In its entirety, there are few numbers that possess the power of 333. Some people are of the opinion that 333 is the weaker version of 666. This means that there is evil present around us, but it is not in its full form. This goes back to our earlier understanding of what the number 666 meant and how it has been used to denote the devil or evil itself in its purest form.

In reality, the number 333 is considered a divine force and even represents the arrival of the angels.

But now the question arises: Why would you need the presence of angels? Is there something major about to happen?

The answer is that yes, there is a big event that is about to occur. The number 333 appears in your life in order to prepare you for what's coming. This way, you are not blindsided by any incident or event. You are prepared to meet it head on and take the necessary action.

If you spot 333, then get ready for something big!

444

This is an unusual number to spot, not because it has no meaning, but it is simply a form of affirmation.

444 is yet another number that is not examined by breaking apart the digits. You understand it as a whole.

When you spot 444, it usually means that you might have experienced something positive in your life recently. It could be something small or something big. Whatever it is, it will continue on for some time.

You might think that 444 seems like the universe is wasting time with numbers. Why tell us that something good is going to continue to happen for a little while longer?

Sometimes, people need to be reminded of the good things in their life. They need to know that they are capable of receiving positive influences. Many people, because they forget to look at the good things, derail their lives. They head into depression or begin to entertain more negative thoughts. They start feeling that they are not capable of having positive influences or occurrences in their life or they might feel that they are unlucky individuals.

But if the universe itself is reminding you that there are indeed good things happening in your life and that you are going to have more of them, then it changes your perspective. It's like your thoughts have taken a complete 180-degree turn. Now you see the good things. You spot the people in your life, the opportunities you have, the good health you have maintained, or a plethora of other positive presence in your life. You suddenly shift your focus to the good rather than the bad. This eventually makes you want to achieve even more.

You could say that 444 is the universe's motivational speech. Go get 'em tiger!

555

This is another number that is taken as a whole rather than the sum of its parts.

If you have been on a spiritual journey recently, then 555 is a sign that you have to continue on that journey. The road ahead might be challenging and your decisions might be tested and rebuked, but you have to focus on the end goal.

Because what you are about to get in the end is going to be truly rewarding and will possibly be a life-changing event. The idea of following your spiritual path is focused on here because perhaps you have chosen this path after many challenges in your life or you would like to make big changes. Because your focus—by choosing the path of spirituality—is to adopt a life that is going to attract more benefits, positivity, and opportunities, the number is encouraging you to continue your spiritual efforts.

After all, the end result is not just about your spirituality. It is about you as a whole.

Number Combinations

Number combinations are far more complex and rare. Remember the idea of how the universe does not reveal all its secrets at once? One of the reasons is because of the complexity of its information.

The language of the universe cannot be expressed through a mere 26 letters. In fact, even if you take the language of Khmer (spoken in Cambodia) which is reported to have the most letters in a language (74), you still cannot compress all the information of the universe into a meaningful sentence. All you can do is try and figure out the clues that are given to you.

Clues that arrive in the form of number combinations.

Let's take a simple example.

What if you start noticing the number 349?

You might notice 3:49 on your clock. You might start seeing that the number repeats itself in many ways. The phone number of the hotel you were going to check into coincidentally starts with 349. You might even get placed in room 349. Perhaps your area code begins with 349. In such ways, you start noticing the number in your life.

What do you do? How can you interpret that?

The first thing you do is break apart the numbers. You now have a 3, 4, and 9.

3 talks about self-expression. It is about discovering your talents and bringing them out into the open.

4 is about your career and work.

9 is about a sense of accomplishment.

Once you have understood the three digits, it is time to start examining your life and see how they represent it. Based on the above meanings, you could come to the following conclusions:

- You might feel that you are not getting enough opportunities to express your talents at work. It might be because you feel that your role is focused on mundane details. As such, you do not feel a sense of accomplishment. Perhaps you might need to look for avenues where you can find more challenges.

- You have recently honed your talents. You feel confident about your capabilities. However, you do not know how you can convert it into work. You are looking for either a job or self-employment opportunities where you can put your talents to practice and earn a little while you are at it. You feel that this will give you a sense of self-accomplishment.

- You are trying to balance your passion and your career. Because of your working hours, you have difficulty managing time. This has eventually caused a disruption in your ability to use your passion to the maximum. Eventually, you feel like you haven't accomplished a lot in the long term.

The above example is just one of the many ways that number combinations might present themselves to you. All you have to do is use this book as a reference and work on each digit. Then think of what meaning they are presenting to you as you examine your life and its challenges.

How Will Numbers Manifest in Your Life?

One of the common ways that you will spot number sequences is by looking at your digital clock or analog watch. You might notice it more precisely when you refer to a digital source for the time, like the aforementioned digital clock or an electronic device like your mobile phone, laptop, or smart television.

Let's take the example of 1111. You can spot the 11:11 time on your watch, but you might not pick up on it. After all, the hour hand will be pointing at the number 11 and the minute hand will be pointed towards the 11th digit in the arrangement of your watch.

As you start your day, you might notice the number repeat itself on your financial transactions. Perhaps you bought a coffee and breakfast at your favorite cafe and the bill comes up to $11.11. You might be checking your monthly bills and the total might be $1,111. If your utility bill comes with a number sequence (which is used as an identification number for the bill), you might notice 1111 present in the sequence.

In similar ways, be on the lookout for the appearance of numbers in numerous positions and scenarios.

Chapter 3: Numerology in Your Life

We have just taken a glimpse into how numerology can affect our lives in different ways. By understanding what each number represents, we are able to find out more about the guidance that they give us.

This time, we are going to take it a step further. We were in the balcony seats when we were examining the numbers. Now we are going to take a front-row view of the whole concept of numerology.

Let us look at some of the aspects of life and how numerology affects them.

We will start off by looking at one of the biggest factors that dictate the way we live our lives: our health.

Numerology for Your Health

Each number from 1 to 9 represents one aspect of your life and dictates what you should be looking for when you see those numbers around you.

Numbers 4 and 5: Physical

- When you get these numbers, then you are considered an individual who has a practical, down-to-earth, and sensible approach to your health. You may be hard-working, but you are the one who works hard by looking after your health as well.

• You like communicating your worries to others. If you are affected by a physical malady, it is not like you to keep it inside yourself. You like to find a solution to your physical problem as soon as possible.

• However, you are the kind of person who gets easily burdened by your physical well-being. Because you are highly conscious of it, you feel its effects impact you greatly. You might also become restless or prone to unpredictable changes when it comes to your physical aspects. When you pick up a new habit, you might drop it suddenly and without any explanation regardless of the consequences such actions cause.

What the Numbers Are Telling You

• If you have experienced physical problems or have been trying to make some changes physically, then the appearance of these numbers means that you have to focus on your change and not give up.

• In other words, make a change but do not become restless about it or drop the change suddenly.

• Here's an example: Let's say that you would like to lose weight. You are fully conscious of how your weight has been affecting you. You decide to head to the gym and sweat it out. Of course, you might not notice the results for a while. This might cause you to become restless and you might be tempted to stop going to the gym. Don't do it!

• Rather, make the change a habit. You are rather expressive, so communicate your intentions to those people around you and let them motivate you as well. You are a person who focuses on your health so allow yourself to keep that in mind as the end goal.

Numbers 2, 3, and 6: Emotional

- You are fairly sensitive, but you use your emotions to help others in a positive way. You are typically loving, warm, and generous. Peace and harmony mean something special to you, and you look for them in your life.

- However, this also means that you can tend to be brooding, self-absorbed, and moody. You sometimes get so carried away by how you feel that you fail to notice the feelings of others. Alternatively, you end up giving too much time to others while failing to take care of yourself.

What the Numbers Are Telling You

- Make sure that you are giving enough time to both yourself and your friends. This does not mean that you have to always be vested emotionally. What it means is that you should focus on enjoying the simple things in life.

- Spend more time with the people close to you. Choose to go out or indulge in a pastime or hobby to distract yourself from dwelling on your emotions too much.

- Watch out! If you continue on your emotional path, you might eventually affect your mental health as well.

Numbers 1 and 8: Mental

- You are a bright individual. You ooze positive energy, and you are enthusiastic about many things in your life, especially your job, your love life, your passions, or your present situation. You like to be leader, or you might already be a leader. Either way, you like to take charge and be in control.

- You may love to engage in intellectual conversations, and you are charming and witty.

- However, you have a tendency to feel insecure.

- You might also become dominating in a scenario or conversation to such an extent where you might not allow the other person to speak or express their opinion.

- You enjoy the spotlight a bit too much. This might make people think that you could be a narcissist when you are actually not.

What the Numbers Are Telling You

- You might be alienating people with your dominating personality. Though you might not narcissistic, it might appear that way because of the way you deal with various scenarios.

- Your insecurities might be causing you to lose wonderful opportunities or perform poorly in your life–whether it is in everyday matters or with other people–or simply with the various projects that you are handling.

Numbers 7 and 9: Spirit

- You are typically shrewd, opinionated, and can hold yourself well in any conversation.

- You are a quick-thinking person who is able to form opinions carefully and without offending anyone. You are typically kind and considerate when you give your opinions to someone else.

- An air of knowledge and wisdom surrounds you, and you use them to be a mentor, counselor, a guide, or a friend.

- However, there are times when you tend to be absent-minded about life. You might miss out on big opportunities

- You can become unrealistic in your approach to life's challenges.

- You suddenly start feeling vulnerable, and this causes you to go into panic mode.

What the Numbers Are Telling You

- Becoming spiritual means that you have a clear sight of the goals you want to achieve. Spirituality should help you focus on your life, people, dreams, goals, and ambitions. You should not get distracted easily. These numbers are letting you know that you might be losing sight of things that are important to you.

- There are chances that you are losing touch with your spirituality or you might not be putting it to good practice.

- Beware of people who are a negative influence in your life.

Numerology in Pregnancy

The birth of a new life is always something that should be celebrated. It is a moment of joy and pure love.

One of the most important questions asked when it comes to numerology is: When is the right time to get pregnant? And what does it mean to get pregnant in a particular year?

The first thing that you have to do is find out your birth number. This is done by taking into account your birthday and adding the numerical value of the digits.

For example, let us say that you were born on August 9, 1987. The method to calculate your birth number is:

$8+9+1+9+8+7 = 42$

$4+2 = 6$

You now have your birth number. We then take the year that you are planning for your or your wife's pregnancy. Let us say that the year chosen is 2020.

You now have to get your pregnancy year number. You get it by:

$2+0+2+0 = 4$

You then add your birth number and your pregnancy year number.

$6+4 = 10$

$1+0 = 1$

The final number that you get will then be used to find out more about your pregnancy based on the table below.

Number	Meaning
1	The pregnancy has better chances of going well, but it might be a challenge to focus on both career and pregnancy. Alternatively, it could also be that your spouse might find it challenging to give attention to both a career and the pregnancy.
2	This year is wonderful for getting pregnant. But if you are planning to work until the final days of the pregnancy, then this might cause more stress on you and the baby than is necessary.
3	This is a neutral year. It does not lean heavily toward a positive or negative outcome.
4	This is not the most ideal year for pregnancy, and you are likely to face problems and changes in your life. However, you can try to withstand these problems and changes with your

	spouse.
5	This year is a blessing if you end up having twins. It is neutral if you don't.
6	This year shows that though you are more likely to avoid having major health complications during pregnancy.
7	This year has a higher chance of causing postnatal depression. If you would still like to get pregnant during this year, try and make sure that you have eliminated any potential causes for stress, mental issues, and emotional attacks.
8	This year is ideal if you are not planning to have twins.
9	This year might have more hospital trips than regular.

With the above table, you will be able to plan your pregnancy better. One piece of advice here is to try and involve your spouse and other members of the family. This is because no matter what decision you make and what challenges you face, being aware of the potential challenges will help you and your spouse or family to plan your pregnancy.

On a completely unrelated side note, if you are indeed planning a pregnancy or if your spouse is, then I wish you all the blessings on your pregnancy. May the numbers and the stars align in your favor!

Numerology for Positive Energy

When you are faced with a constant flow of negative energy and you do not know how to deal with it, then the power of numerology can bring balance into your life.

To understand what you must do, your first step is to focus on finding out your Fibonacci number.

We are going to take an example here, and for the sake of the example, let us assume that your Fibonacci number is 3.

Once we have that number, we are going to focus on the present year. The important reason for considering the year is because what you face this year might be different from what you face next year. Even if you face the same challenge next year, the incidents and situations surrounding the challenge might be different.

For now, let us assume that it is the year 2020. The resulting number is:

$2+0+2+0 = 4$

Now combine your Fibonacci number with the year number.

$3+4 = 7$

Using that number, find out how you can attract more positive energy into your life form the table below:

Number	Attracting Positive Energy
1	Focus on exercises Get into the habit of reading a book
2	Build positive relationships Help other people out (you could become part of a charity drive or volunteer at a local non-profit organization such as animal welfare institutions, orphanages, food drive camps, or others)
3	Adopt any creative activity such as painting, baking, writing, music, or any other artistic endeavor
4	Change your diet to healthy meals Take more walks with your friends
5	Focus on physical activities Get a massage!
6	Focus on your appearance and try

	to groom yourself more Spend a little on getting new clothes, accessories, or cosmetics
7	Avoid isolation Take up meditation
8	Spend time outdoors Perform tasks outside such as gardening, walking your dog, or other such activities
9	Maintain a diary Spend time with friends or family Adopt a pet if you can!

Numerology for Success

How successful you are, depends on the people who are there to help you reach the top. This does not mean that hard work and talents are not important. However, hard work, skills, knowledge, creativity, and talents cannot be measured. The degree to which you use them depends on the work that you are doing and what the task at hand requires from you. For that reason, let us look at the many ways that you can use numerology to find the right person to help you with your success.

This person you are looking for can be a spouse, business partner, someone who can help you get a loan, an investor, a family member, or anybody else. You can measure success in almost any sphere of your life, from your health to your love life to your relationship with your parents. For that reason, we are looking at ways numerology can help you attain your success.

Let us take a situation to help you understand how success works. Let us assume that you have been on bad terms with your parents. You have stayed silent with them for a week now, and you feel that you have to take the next step in moving your relationship back to the way it was. You decide that the best way to go forward is by first talking to someone before going to talk with your parents.

You narrow down the list of people you would like to have a talk with to two individuals.

Your friend, Bruce Banner.

Your sister, Natasha Romanov.

We are going to once again use the Fibonacci table.

1	2	3	4	5	6	7	8	9
A	B	C	D	E	F	G	H	I
J	K	L	M	N	O	P	Q	R
S	T	U	V	W	X	Y	Z	

Let us assume that your Fibonacci number is 1.

We are now going to work on the Fibonacci number for Bruce Banner.

B+R+U+C+E+B+A+N+N+E+R

2+9+3+3+5+2+1+5+5+5+9 = 49

4+ 9 = 13

1+3 = 4

Next, we focus on Natasha Romanov.

N+A+T+A+S+H+A+R+O+M+A+N+O+V

5+1+2+1+1+8+1+9+6+4+1+5+6+4 = 54

5+4 = 9

Now, you then refer to the table to see the compatibility number. Basically, this is the number that lets you know who can help you achieve the success you desire.

Number	Compatibility Fibonacci Number
1	Those with 1, 4, or 8 as their numbers will help you achieve success. People with 2 and 7 could also help you, but their help might be short-lived.
2	Those with the numbers 1, 3, 4, and 8 are more likely to help you toward success.

3	Those with the numbers 2 and 9 are more likely to help you toward success.
4	Those with the numbers 1 and 8 are more likely to help you toward success.
5	Those with the numbers 1 and 9 are more likely to help you toward success.
6	Those with the numbers 3, 6, and 9 are more likely to help you toward success.
7	Those with the numbers 1, 2, 4 and 7 are more likely to help you toward success.
8	Those with the numbers 1 and 4 are more likely to help you toward success.
9	Those with the numbers 2, 3, 5, 6, and 9 are more likely to help you toward success.

When you look at the table above, then you realize that by talking to Bruce Banner, you are more likely to feel better, gain some wonderful tips, and even have his support all the way through. Your chances of success with Bruce increase dramatically.

Let's take another example.

You have been approached by two partners to start a business. On one hand, you have Thomas Cook and on the other, you have Francis Drake.

After you have checked all their credentials and qualifications, you are still at a loss as to whom to choose. At this point, it feels like it is a matter of a coin toss to figure out who should be your partner.

But rather than leaving your odds to a coin, why not find out for yourself?

Let's say that right now, your Fibonacci number is 5.

After using the table, you find out Thomas Cook's number, which is 3.

Using the table again, you find out the number for Francis Drake, which is 1.

From the table, you know that people with the Fibonacci numbers of 1 and 9 are more likely to help you attain success. You now know with whom you should partner with. Time to welcome Francis Drake to the team.

Numerology for Relationships

When you are focused on relationships, it is all about the other person. When you understand the other person's number and how it influences their behavior, you will have a better chance of developing a stronger relationship with them. Whether the person is your parent, spouse, child, friend, or anybody else who is part of your life, you can use the numerology guide to forge stronger bonds or strengthen the bonds already existing.

Here is a table that helps you understand people with different numbers.

Number	Personality
1	This person enjoys being in charge. They value independence and taking control of the situation. They know what they want and mostly make their own decisions. Do not try to fix or change them. You might only end up pushing them away. Being needy or nagging them definitely turns them off. If you would like to communicate something to them, never take an indirect approach. Always be straightforward with them.
2	People with 2 as their number love the very idea of being in love. They value companionship and intimacy. They love to be with people, whether it's their partners, friends, parents, or children. Quiet and peace are something they value most as they can be quite sensitive. Do not let their need for quiet push you away as they always seek to have you close to them. By being sensitive, they go the extra mile for the people close to them. But this has the tendency to cause them emotional pain on a deeper level if they are hurt or attacked.
3	Those with 3 are deep thinkers and sensitive as well. They are social butterflies and enjoy meeting new people, spending time with their

	friends, or simply hanging out at social events. People feel pleasant around them as they engage and interact well with others. Because they love to create, you can find them making something from scratch, whether it is a new project or a hobby. Optimism comes naturally for them, and they love to make people smile.
4	Strength, devotion, and loyalty are traits that are valued by people with the number 4. They want to take care of you and feel protective of you. Doing things for you is something that they enjoy. However, with all of this attention, they are down-to-earth people. With logic being more important than feelings, you cannot expect them to do something as spontaneous as simply dropping everything and taking a vacation!
5	People with the number 5 are creative and witty. They keep the atmosphere lively and avoid dull moments. Their adventurous spirit helps liven up the moment, and they are mostly looking for things to excite them. They are searching, exploring, and finding new things to engage with and experience.
6	People with the number 6 are highly responsible people. They also enjoy being with someone they care about. This does not mean that they are independent. Rather, they love to focus their attention on people. They are the ones

	who are looking out for the stray dogs or cats in the neighborhood. They want to shower love and adoration on people, but they are particular about who those people are.
7	The number 7 focuses on people who love to have meaningful conversations, explore deep topics, and crave knowledge. People who come across them often say that they have a certain otherworldly energy about them. They are intuitive and empathetic. You hardly have to clue them in on what you are feeling. They can easily interpret your emotions.
8	8 is the number for people who are charismatic. They love to succeed in life and are mostly seeking the next venture to work with. They love making things grand, no matter what they are dealing with. They have an aura about them that shines bright and sucks you into that brightness. You cannot help but feel at ease or joy around an 8. They can also be great motivators to help you believe in yourself.
9	People with number 9 look at things from a unique perspective. They read between the lines and are deeply introspective. This does not mean that they are aloof. On the contrary, they understand themselves and others well. They enjoy doing work for humanity. In fact, they take the initiative in humanitarian causes. They are

	creative, fun, and try their best to look at the world through rose-tinted glasses.

Let's say that you have named your son Jon Snow. When you look at his Fibonacci number, you realize that it comes to 2. Here are some ways you can build your relationship with Jon Snow:

- Spend more time with him. It does not matter if you have something planned or not. The very idea that you are with him makes a big difference.

- When he does something for you, appreciate him with genuine warmth. He is sensitive, so if you give him a half-hearted response, it might hurt his feelings.

- When he is in a bad mood, give him some space and quiet. And when he is ready to talk, listen with honesty.

- Have a special day where you take him out for an activity.

- Do not be afraid to show that you care for him. Even a simple, "I love you, son," while you are putting him to bed has a deep effect on him.

In such ways, simply finding out the Fibonacci number will help you find ways to improve relationships with people.

Chapter 4: The Beginnings of Astrology

Astrology is the practice of reading the patterns and movements of the celestial bodies that include the planets to find a correlation between them and your life. Astrology believes in the idea that the universe is interconnected in ways that we cannot yet fathom.

People might think that astrology is irrelevant in today's world. They believe that the practice of reading the stars is archaic and has no benefit to people living in the present.

The reality is that astrology has found new importance among the people living in the 21st century. People are not satisfied with the common knowledge about the connection we have with the cosmos at large.

Think about this: Everyone is made of star-stuff.

The Big Picture

Sometimes, when two people meet each other, one of the common questions asked is what the other person's star sign is. They are fascinated by astrology, yet their perception is narrowed down to the horoscope columns in magazines or websites. While those columns are indeed the final product, there is a much bigger story behind those horoscopes.

And this is a story written in the stars.

You, me, your neighbor, Ryan Reynolds, Joan of Arc, the chef of the restaurant you love so much, and everyone on this planet is made because stars exploded in the distant past. The explosion of the stars created new elements. In turn, new stars began to form and newer elements were created. Eventually, these elements combined together to form the most basic forms of protozoa millions of years ago.

Today, those elements have evolved to become beings who are capable of exploring new ideas, binge-watching *Game of Thrones* in one sitting, exploring the stars, helping each other, and creating wonderful things.

Those elements have created you, the person who is reading this book and is hopefully finding it spectacular. They have created me, who loved writing this book for you.

Astrology understands this connection that we have with the celestial bodies (after all, they were created from the same star stuff). Their movements and positions in the universe talk of an ancient story, one that reveals the tales that are weaved in our lives.

Astrology Through the Ages

When one thinks of astrology, one might not think it has any history.

But the opposite is true.

In fact, you can trace astrology back to the world of ancient Babylon and from there, to Egypt and Greece. During those times, Alexandria was considered to be the knowledge center of the world. After all, one of the biggest libraries to have existed in the world existed in Alexandria. Great minds would flock to the library to learn, debate, research, and make monumental discoveries. Whether it was physics or mathematics, philosophy or religion, one could find a vast repository of knowledge in The Great Library of Alexandria. It birthed great minds like Eratosthenes, who was the first person in the world to discover that the Earth was round. The famous mathematician Euclid–who is considered to be one of the most important mathematicians ever–took many research materials from the library.

In similar ways, a great many names passed through the halls and a great many subjects were developed.

Subjects like astrology.

Back then, astrology took inspiration from various sources. From the Greek ideas of the four elements of nature to the Babylonian concept of the zodiac to even the symbolic interpretations of Egypt. Astrology was complex and was not a result of someone's hobby. It was a field of study that encouraged scholars, thinkers, and philosophers to ponder about its mysteries. In fact, philosophy was often combined with astrology to give more meaning to the universe we live in.

Eventually, astrology became important to the Romans in the 2nd century BC. But alas, the fate of the Great Library of Alexandria would be one of fire. Literally.

After the fall of the great library, astrology came alive through Islamic scholars who worked to keep the field alive by documenting it. Furthermore, they continued to develop the field through their own discoveries.

The 12th century brought about a renewed interest in the field, and the studies done by the Islamic scholars were translated into Latin. From there, astrology spread to different parts of the world before being available to practically everyone on the planet.

Today, even the millennial generation is finding more interest in astrology. This is mainly due to the level of accessibility one has to the subject. Simply boot up your favorite internet browser and do your own research on the subject. You can find thousands upon thousands of sources that will connect you to astrology and its predictions.

Astrology and Your Life

One of the misconceptions people have about astrology is that one should 'believe' in it.

But the reality is not of belief. One does not usually 'believe' in biology, chemistry, or psychology. Sure, one can believe the results and information the subjects provide, but that belief talks about the confidence placed in the information received.

The same case applies to astrology. It is not a religion. It provides you with insights and understandings about us, our lives, and our connection to the grand cosmos.

You are not required to worship any aspect of it or revere its teachings. It guides you when you are looking for guidance. But when you are engaged with your day or in some activity, it does not ask you to silently contemplate it and thank it.

In short, astrology helps guide your life when you feel lost in your challenges, paths, and choices.

It is true that astrology studies writing in the skies. It focuses on the motions of the celestial bodies and cosmic events. But it does so to give more insights into the life here on Earth. It is about us. But as we saw earlier, we are connected to the trillions of cosmic materials and bodies out there. If there are answers, then they are found by looking at the bigger picture. Think of astrology like opening a big book about our lives and looking at all the chapters within it. Without it, we are probably focused on one page or probably a single paragraph, hoping to make sense of the information given there. But astrology reveals all the pages and then lets us decide what we could use to make our lives better.

Astrology and Psychology

It might be rather odd to put those two words in the same sentence, but there is a reason.

Most of the time, we hope that life runs smoothly and rarely presents us with the occasional bumps in the road. During such times, we have things handled pretty well.

Other times, however, we are faced with crises and uncertainties. It becomes useful, then, to have a way to look at these situations and understand what is happening from a much deeper perspective.

When astrology speaks to us through its symbols and meanings, psychology acts as the interpreter. Astrology does not simply throw in some facts and be done with it. It asks you to reflect on its messages. More importantly, it encourages you to seek out self-reflection and self-knowledge. However, focusing on the self is not easy to do. Our minds are far more complex than we understand, which is where psychology comes into play.

Think of astrology as the wise, smart, and intellectual ambassador of a far-off nation. The ambassador has arrived to bring news about the many events that could affect the way we live. But no matter what the ambassador says, the information provided talks about scenarios, events, and factors in our lives that might require a bit of an explanation. That's where psychology comes in. It acts as the interpreter, allowing us to recognize how our lives are affected and what we can do with the information provided.

However, this does not mean that astrology is confusing. Rather, astrology is about personal growth and understanding. And nothing helps figure out personal growth better than psychology.

Chapter 5: Astrology and the Planets

If you think of astrology as a chessboard, then the planets are the game pieces. Every move they make leads to a different outcome.

Through the eyes of psychology, you can consider planets as a representation of many principles that exist in this world and within you. This is why the connection between the planets and you explains your personality.

At this point, you might think to yourself, "I get what you are saying, but I think you are making a grave mistake. You do realize that the sun and the moon are not planets right, dear author?"

I do. However, in the world of astrology, the sun and the moon are referred to as 'planets' as well. This is not for any symbolic meaning, but more for brevity. It confuses people when someone says celestial bodies because then they think that the stars are included as well (the stars are, but for now, we are only focusing on the planets). Alternatively, rather than saying, "the planets, the sun, and the moon," it is much easier to simply say the planets as a broader term.

Let us now dive further into the planets. An important note to be made here is that Uranus, Neptune, and Pluto are called 'collective' planets. This means that their meanings are not very personal like the other planets. You get to understand the overall theme of their meanings and then extract personal meaning from that. Which is why the best way to explain their overall meanings is by comparing them to important moments in history. This allows you to look at that moment in history or time, understand the overall lesson or meaning one can derive from that moment, and then discover how they relate to each sign.

The Sun: Identity and Life Purpose

Characteristics: Vitality, purpose, self-development, inner authority, self-confidence, creative power, masculine

Planetary Day: Sunday

Rules: Leo

Colors: Orange, yellow, and gold

Metal: Gold

Exploring the Sun Through the Signs

Aries: You may be called to be a leader or a pioneer. During such times, you will need to awaken your warrior spirit.

Libra: The idea of balance and harmony is important to you. You seek them out through the relationships you have formed or the expression of your artistic side.

Taurus: The gift of relationships means something significant to you. It is for this reason that you aim to create something that is enduring with the people you care about.

Scorpio: You value the journey taken that involves numerous challenges and difficult conditions. The transformative powers are what you seek.

Gemini: Knowledge is something you crave. You also look to learn languages and improve your communication skills.

Sagittarius: You enjoy exploration. Traveling is something you are fond of, and you try and make it a point to visit as many places in the world as possible.

Cancer: Taking care of a family or improving the lives of a group of people close to you brings you joy and warmth. Essentially, you love to be the caretaker.

Capricorn: Life has been placing you in the position of a leader for quite a while. Eventually, your leadership qualities will help you with self-reliance.

Leo: When you aim to accomplish something, you wish to be admired for it. You are not afraid of being a leader, but you would like people to recognize your work.

Aquarius: Social equality means something to you. You uphold the principles of equality in your life, as well, which eventually transforms into your social identity.

Virgo: Developing a craft lets you feel a sense of accomplishment. Ideally, you aim to use your craft on a professional basis and improve your skills.

Pisces: You enjoy using your powers of imagination to create wonderful results. There is a sense of achievement and pride when you complete a creative project.

The Moon: Protection

Characteristics: Instinct, nurture, security, feminine, emotional, physical, maternal, intuition

Planetary Day: Monday

Rules: Cancer

Colors: Silver and white

Metal: Silver

Exploring the Moon Through the Signs

Aries: It is not beneficial to anyone when you are held back and prevented from achieving your potential. You also encourage others to become independent.

Libra: The idea of being at home with the people you love appeals to you more than being by yourself. However, you do crave periods of peace and calm.

Taurus: Being methodical is how you approach things. The idea is that you like to take things calmly and slowly, experiencing them properly.

Scorpio: Holding on to emotions can be destructive, yet that is what you tend to do sometimes. If you would like to exhaust your emotions, physical activities help.

Gemini: Learning from verbal communication helps you understand people and the world around you. You seek to have a pleasant conversation as much as possible.

Sagittarius: You need more personal space. You like to have room to move around and think. Confined spaces do not work well for you.

Cancer: Security is foremost on your mind. This security could be about your job, your relationships, or your future.

Capricorn: When you have structure and routine in your life, you excel at many tasks. Chaos tends to cause more confusion in your mind than is necessary.

Leo: You are looking to receive love and praise from the people around you. Your talents develop themselves naturally.

Aquarius: Community is something you aim to be part of. Along with your family, you look to maintain good relationships with your neighbors as well.

Virgo: Whenever you take care of people close to you, there is a practical method to the way you approach things. You like to think things through logically.

Pisces: By being strongly empathetic, you are sensitive to the feelings of others. This might also cause you emotional pressure if you cannot control it.

Mercury: Learning

Characteristics: Learning, education, opinions, trade, youth, language, exchange, communication, mind, thoughts

Planetary Day: Wednesday

Rules: Virgo and Gemini

Colors: Grey

Metal: Mercury

Exploring Mercury Through the Signs

Aries: With the power of Mercury, you become better at making decisions and thinking clearly under pressure.

Libra: This planet gives you the ability to be tactful, while also appearing charming and engaging. You also get to improve your powers of observation.

Taurus: Your thoughts are more methodical and considerate. When your thoughts are not in a hurry, you are able to face challenges better.

Scorpio: You are emboldened to speak your thoughts. At the same time, you become inquisitive about the motivations of others.

Gemini: Curiosity guides you under this planet. Your sense of wit also increases, making you more sarcastic than normal. Your curiosity to learn increases as well.

Sagittarius: You develop a sense of open-mindedness about the world around you. There is a yearning within you to look at the broader aspects of the world.

Cancer: Memory becomes affected by this planet but in a good way. The powers of retention and recollection improve, giving you better capabilities to learn things.

Capricorn: Multitasking becomes easy. The ability to juggle different tasks becomes a walk in the park for you. At the same time, you are curious about setting bigger goals.

Leo: Your confidence is raised during this period allowing you to seek out activities that challenge you to step out of your comfort zones.

80

Aquarius: The many aspects of the world and their scientific reasonings become something you aim to look for. You also navigate toward debates and discussions.

Virgo: You feel the need to indulge in activities that allow you to use practical skills, such as DIY crafts, woodwork, and other such activities.

Pisces: You become curious about exploring your imagination and putting it to use. You want to write stories through words, pictures, or other forms of media.

Venus: Love

Characteristics: Romance, relationships, beauty, pleasure, aesthetics, sexual desire, erotic love, enjoyment

Planetary Day: Friday

Rules: Libra and Taurus

Colors: Green

Metal: Copper

Exploring Venus Through the Signs

Aries: You will not wait around to be taken on a date. You want to be chased by others or shown attention by the other person.

Libra: An increase in your charm is what you are going to experience with Venus. But at the same time, you develop a sense of gracefulness about you.

Taurus: Love becomes something you want to experience through your senses. Activities such as long walks, wonderful dinners, and visiting art galleries appeal to you.

Scorpio: With both feet in the air, you fall in love deeply. You do not take half measures. You want to be in love and expect the other person to feel the same.

Gemini: When in love, you begin to lay down the foundations for good communication. You like your partner to speak openly, as you are planning to do the same.

Sagittarius: Free bird! That's what you are. To you, love is an adventure. For this reason, you might be looking for a partner who can travel with you.

Cancer: While the physical aspect of love is important, you value the emotional bonds that you develop with people. You aim to build a family.

Capricorn: You begin to value structure. This means that you don't like surprises in your love. You believe in setting boundaries that both parties can respect.

Leo: Impression is the name of the game. You want to create lasting impressions on the people you want to develop romantic relationships with.

Aquarius: You value space, and you are looking for love that does not need words to keep it going. You don't mind finding someone who enjoys the quiet moments.

Virgo: Love means acts of kindness and generosity. This is what you look for every day, even though those acts do not have to be something big.

Pisces: Your goal for when you find love: devotion. Honestly and loyalty are things you value above all else. You don't stand for anybody breaking your trust.

Mars: Action

Characteristics: Strength, survival, anger, energy, competition, conflict, vitality, survival, courage, daring

Planetary Day: Tuesday

Rules: Scorpio and Aries

Colors: Red

Metal: Iron

Exploring Mars Through the Signs

Aries: You want to channel your energy into your life. At this point, you would like to make some important actions.

Libra: When you want to be assertive, you do so without coming off as aggressive and threatening. This might be an advantage that you might use.

Taurus: You are not as quick to anger as the other signs. Only through constant provocation does your anger manifest itself.

Scorpio: Endurance is in your blood. You have the drive to survive the most extreme situations. At this time, you prefer to act covertly in many aspects of your life.

Gemini: Defense and attack maneuvers come in the form of words. You use verbal shots to make your point, but with little trace of anger.

Sagittarius: The great outdoors begins to appeal to you. You have the energy to venture outside and avoid spending too much time inside.

Cancer: You try and avoid conflicts. However, you are not afraid of facing them with renewed vitality if your loved ones come to harm.

Capricorn: There is a sense of energy that compels you to pick your battles and challenges. You want to keep your energy focused rather than waste it on anything.

Leo: To you, energy comes in the form of athleticism. You take pride in your athletic ability and would like to explore how you can use it.

Aquarius: When you are challenging social injustices, you have the courage to voice your opinions, but you maintain a cool head at the same time.

Virgo: Building, doing crafts, or fixing things help you keep your anger in check. You usually tend to avoid conflict but you are ready to defend yourself if need be.

Pisces: You try to avoid conflicts and tough situations until they blow over. You take the time to plot your next strategy.

Jupiter: Growth

Characteristics: Possibilities, blessings, enthusiasm, morals, ethics, opportunities, fortune, abundance, adventure

Planetary Day: Thursday

Rules: Pisces and Sagittarius

Colors: Purple

Metal: Tin

Exploring Jupiter Through the Signs

Aries: Using your vision, you have a tremendous capacity to motivate others. You can inspire people to achieve more.

Libra: The growth that comes into your life might be possible through a two-way exchange. This means that you are going to be using help from someone.

Taurus: To you, abundance is achieved by being patient in achieving your goals. While your goals may be lofty, your mind is steadfast.

Scorpio: You are passionate about your goals. You carry forward using enthusiasm, and it is this feeling that propels you to face challenges without changing mind.

Gemini: There is a deep need to know more about the world. You feel that there is growth in knowledge.

Sagittarius: With the presence of Jupiter, you feel as though you have to gain a wide variety of experiences to improve in what you want to do.

Cancer: A sense of working with other people develops within you. You find possibilities with people, and you want to work with them toward a goal.

Capricorn: You have a sense to nurture your leadership abilities. You also have a hunger to look for better opportunities.

Leo: Whatever you do, you want to do it with a sense of grandness. This allows you to see possibilities when there do not seem to be any.

Aquarius: Optimism drives your vision. You are willing to get where you want to without passion. But you have a strong moral code, as well.

Virgo: You pay attention to details. You feel that there is good fortune by looking out for opportunities in all areas.

Pisces: Using good ethics, you want to reach a better place than where you were in the past. You want to prove to others that there is a better version of you for them to see.

Saturn: Maturity

Characteristics: Authority, discipline, dignity, responsibility, rules, tradition, limits, concentration, patience

Planetary Day: Saturday

Rules: Capricorn and Aquarius

Colors: Dark brown and black

Metal: Lead

Exploring Saturn Through the Signs

Aries: Using discipline, you create a sense of organization in your life. You have the strength and courage to work toward what you want.

Libra: You gain a sense of impartiality. Even when you are an authority figure, you ensure that everyone is treated the same.

Taurus: You like to follow the rules because you aim to create a sense of order in your life. You do not venture forth to break the rules too much.

Scorpio: You enjoy tradition and the values that it brings. You feel it is important to incorporate tradition into your life and encourage others to do so as well.

Gemini: You develop a sense of dignity. You want to reach a position where you can walk with your head held high.

Sagittarius: Faith may be a matter of perspective. There are limits to how much faith you can put into things. But that does not mean you aim to cause any chaos.

Cancer: You feel a sense of responsibility toward those close to you. You want to do more for them, yet you do not want to hold them back with limitations.

Capricorn: Discipline is important to you. But you are not overbearing. You have the patience to teach discipline to others.

Leo: Whatever you do, you want to do it with a sense of grandness. This allows you to see possibilities when there do not seem to be any.

Aquarius: Your concentration is focused more on creative pursuits. You want to achieve big things in the creative endeavors that you have taken.

Virgo: You have the patience to excel in the field of your choice. But you often feel as though you are held back by certain rules.

Pisces: Placing limitations on yourself might be difficult for you. However, you do understand that sometimes, there is a reason for those limitations.

Uranus: Change

Characteristics: Rebelliousness, intellect, originality, independent, radical, revolution, innovation, free

Planetary Day: No specific day

Rules: Aquarius

Colors: Electric blue

Metal: Uranium

Exploring Uranus Through the Signs

Aries: Change in history that defines you: the pioneering of the production line to manufacture more cars. You want freedom in many areas and you aim to get more of it.

Libra: Change in history that defines you: the equality of gender becomes an issue. You are aiming to bring about a sense of balance in your life.

Taurus: Change in history that defines you: the change in farming methods due to the Great Depression. You want to find alternate ways, even when the going gets tough.

Scorpio: Change in history that defines you: the spirit of the age expressed through the Punk movement. You look to find more ways to express yourself.

Gemini: Change in history that defines you: the revolution in communication through the invention of radar. You plan to improve communication with people.

Sagittarius: Change in history that defines you: the freedom to explore the world through the discovery of air travel. You want to explore the world on your own.

Cancer: Change in history that defines you: the focus on rebuilding homes and communities after the Word Wars. You aim to rebuild broken relationships.

Capricorn: Change in history that defines you: the change to professional structures due to economic recession. You want to challenge established institutions.

Leo: Change in history that defines you: the way people experience leisure from the invention of television. You want to find new ways to add more adventure to your life.

Aquarius: Change in history that defines you: the innovation brought about by computing. You aim to be more innovative in various aspects of your life.

Virgo: Change in history that defines you: the presence of social change due to birth control pills. You do not want to accept limitations and aim to have better choices.

Pisces: Change in history that defines you: the revolution of the arts due to technology. You want to show people that you have done something new.

Neptune: Imagination

Characteristics: Spiritualism, fantasy, sacrifice, escapism, transcendence, idealism, imagination, dreams, bliss

Planetary Day: No specific day

Rules: Pisces

Colors: Sea green

Metal: Neptunium

Exploring Neptune Through the Signs

Aries: Transcendent moments in history: the birth of modern art. At this point, you would like to explore your artistic side and see what you can make from it.

Libra: Transcendent moments in history: the peace after World War II. There is a lot of chaos in your life, and you want to gain a reprieve from it.

Taurus: Transcendent moments in history: the fall of financial stability. You do not want to depend on only one source of income. You want to achieve more.

Scorpio: Transcendent moments in history: the musical revolution of the 60s. You are not satisfied with how things are. You want to create something unique.

Gemini: Transcendent moments in history: the creation of motion pictures. There is a desire within you to do something creative whether it is small or big.

Sagittarius: Transcendent moments in history: the rise of spirituality. You yearn for more peace and harmony in your life.

Cancer: Transcendent moments in history: the dissolution of old European empires. You want to build something better in your life, and you want to do it with people.

Capricorn: Transcendent moments in history: the end of communism. You feel restricted by the way of thinking around you. You aim to create a social change.

Leo: Transcendent moments in history: the beginnings of cinema. You want to achieve something big, but you want people to see your achievements.

Aquarius: Transcendent moments in history: the innovation brought about by computing. You aim to be more innovative in various aspects of your life.

Virgo: Transcendent moments in history: the growth of

conservative fashion. Even though you are happy where you are, you look for more order in things.

Pisces: Transcendent moments in history: the belief in science. You have a sudden shift toward achieving things by using logical and rational thought.

Pluto: Power

Characteristics: Transformation, control, crisis, change, secrets, survival, renewal, birth, death, metamorphosis

Planetary Day: No specific day

Rules: Scorpio

Colors: Maroon

Metal: Plutonium

Exploring Pluto Through the Signs

Aries: Presence of Pluto in history: the beginning of the Victorian Age. You feel like there is a need for change in your life. One that brings transformation.

Libra: Presence of Pluto in history: the Cold War negotiations. With all the conflicts in your life, you want the birth of better or new relationships.

92

Taurus: Presence of Pluto in history: the change in the American Civil War. There are a lot of challenges in your life, and you are looking to change that.

Scorpio: Presence of Pluto in history: the focus on AIDS. You feel that there are certain things in the past that you have to address.

Gemini: Presence of Pluto in history: the spread of the telephone. There is a greater need to be in better communication with the people around you to create renewal.

Sagittarius: Presence of Pluto in history: the presence of religious intolerance. There is a crisis of conflict within you. You want to challenge established traditions and norms.

Cancer: Presence of Pluto in history: the start of World War I. You know that you are going to face many challenges. But you are determined to survive them all.

Capricorn: Presence of Pluto in history: the economic collapse. You feel like you would like to be financially independent.

Leo: Presence of Pluto in history: the rise of Stalin. You feel like your life is not in your control, and you want to change that.

Aquarius: Presence of Pluto in history: the discovery of Uranus. You know that there are more opportunities for you out there that can help transform your life.

Virgo: Presence of Pluto in history: the start of the environment movement. You want to create a metamorphosis in society or in the world at large.

Pisces: Presence of Pluto in history: the Romantic Peric You are good at what you do, but you feel like there is death of creativity that you can explore.

Chiron: Healing

NOTE: Chiron has often been shifting its status from a comet to a large asteroid to a minor planet. However, in astrology, Chiron is considered a planet because of the way it represents various factors of our lives.

Characteristics: Compassion, mentor, alienation, maverick, teacher, alternative, healer, displacement

Planetary Day: No specific day

Rules: None

Colors: None

Metal: None

Exploring Chiron Through the Signs

Aries: You want to be assertive, but you want to do it in a way where you are thought of as a teacher. After all, you have incredible teaching abilities.

Libra: You are looking for alternatives to many aspects of your life. There is a sense of displacement that you cannot seem to get rid of.

Taurus: There is a desire for you to be known as the maverick in what you do, whether it is in your personal or professional life.

Scorpio: You find great joy in mentorship, partly because it allows ᵥple to see your experience and partly because you get to help others.

Gemini: You are not satisfied with the status quo. You want to achieve more and are looking for alternative paths to success.

Sagittarius: Sacrifice has been something you have been doing lately to get to your dreams. And now, you are looking for a little healing.

Cancer: There is a growing sense of compassion within you. But you want to share that compassion with someone.

Capricorn: You feel like you are the only one looking at the bigger picture among the people you are with. This brings about a sense of alienation within you.

Leo: For too long, you have not been allowed to shine. But it is time for you to show that you can be a maverick as well.

Aquarius: The world needs more teachers. You want to teach something useful to someone, even if it means a little skill to someone you know.

Virgo: Perfection is what you desire, but often, you feel like you are falling short of your target. You are constantly looking for new opportunities.

Pisces: There is a sense of loss within you. This loss could be in your personal, professional, or emotional space. But you are going to find a way to heal yourself.

Chapter 6: Understanding the Zodiac

What differentiates the planets that we just talked about from the signs that we are about to look at?

In astrology, the planets focus on a particular drive. For example, Venus focused on love while Pluto was about power. The planets revealed to you how these drives or elements manifested themselves in your lives. What does healing mean to you? How can the powers of imagination manifest themselves in your life? What do you consider to be change? Each question brings about a certain element that either affects your life or creates a cause.

On the other hand, signs are an expression. They talk about the way various factors in your life, along with the planets, express themselves.

Most of us can easily find out our zodiac signs. All we need is the date of our birth, and by using that date, we can find a corresponding sign. The easiest way to imagine the arrangement of the signs is in the form of a chart because it is more convenient to refer to a chart. In fact, wherever you refer to astrology, you find them arranged in grids, rows, columns, charts, and other similar arrangements. But the most accurate way of looking at the signs is in the form of a circle. The sun is at the center of this circle, and surrounding it are all the signs of the zodiac.

There are many reasons for the circle arrangement:

- The sun, being at the center, sheds light on all your zodiac signs, illuminating them for you.

- The use of a circle speaks of the cycle of the sun around our Earth. Bear in mind that this does not mean that the Earth is the center of our solar system. It's merely the fact that as the Earth rotates, it seems like the sun and the stars are going around it.

Because of that, it almost looks like the sun is making a cycle around our planet.

- Seasons on Earth are mentioned as cycles and for that reason, the circuit of the sun begins in the spring equinox. This does not have any bearing on the actual calendar, but rather on the aspect of astrology. Based on astrology, the sun initiates its journey from the Aries sign and finishes the cycle at Pisces. This fits perfectly with the nature of the signs as well. The sign of Aries is energetic and focuses on commencement. On the other hand, Pisces includes traits such as dispersion and dissolution.

Now that we have understood the basics, let us dive deeper into understanding the signs.

Interpretation of Each Sign

Each of the star signs will provide you with their basic information. This includes the period for which they are assigned. For example, Aries rules the period from March 21 to April 20.

You will then be presented with additional information about the sign itself showing what planet rules it, its element, and it's color.

Once you go through the initial information, you are then presented with the traits of the sign. One thing to bear in mind is that none of the traits are either positive or negative. They simply are. Depending on the situation, they can be used for your benefit or they might cause harm to you.

However, there is a special section that displays the negative traits of the sign. Remember that these negative traits are not like permanent tattoos; they don't define you forever. They are there to show you certain features of your character that you find difficult to change. However, by knowing about them, you can manage these traits and prevent them from getting out of control.

You are also going to learn about their correspondences and compatibilities, which will be explained to you further in the next two sections. Both factors help you understand more about the sign that you are working with.

Correspondence of Each Sign

When we look at the signs, we often think of certain ways to understand them. There are certain trigger words associated with the signs that jolt our memory about information about these signs.

These trigger words are often considered as analogies for the signs, and they are called 'correspondences.'

There is not a lot of information on how these correspondences came to be, but they are all part of the astrological tradition and lore. You might even see snippets of them whenever you are reading astrology in your favorite magazine or online platform. They usually appear as the sign's color attribute or its metal correlation.

The use of correspondences is more than just passing interest. They are used for various reasons:

- They create connections in our minds with the zodiac sign. For example, thinking of the element fire might conjure up all the properties of Aries.

- Furthermore, each correspondence will start to make sense as we get to know more about the sign itself. Of course, each person's correspondence makes sense to him or her only. This is because these correspondences are attributed to the person's traits. For example, Gemini is represented by air, which is symbolic of the way things flow in the world. Perhaps it could indicate the way the individual, like air, has been flowing through life without making any remarkable changes. Alternatively, air flows over all things, so perhaps it could represent the person's ability to examine things from a broader perspective. Each of the correspondences has a meaning that is unique to the individual of that particular sign.

Compatibility of Each Sign

Successful relationships depend on compatibility to a great degree. Typically, we are talking about relationships that are romantic or that involve some sort of partnership, such as friends, business partners, or acquaintances.

But we can also use the charts to check our compatibility with our family.

However, there is an important question to be asked here.

What if we discover that we are not compatible with certain members of our family? Does that mean we have to sever ties with them entirely?

Compatibility can be used for two broad forms of relationships. These are discovery and existing relationships. Here is how zodiac compatibility helps us with these relationships:

- Discovery relationships are those that haven't entered your life yet and you want to include in your life. Whether it is about looking for a new romantic partner or a BFF, you are seeking relationships. Using compatibility, you can find out the people you can easily get along with (compatible) and those who require more work to form good relationships with (non-compatible). Sometimes, people who are not compatible with you might not get along with you at all or refuse to form any connection with you. Either way, you get to choose what you want to do with the information presented to you.

- If you are already in a relationship with someone, whether it is through familial bonds, friendships, professional relationships, or romantic ties, then you can find out your next course of action. When you have knowledge about the other person, then you can find ways to work with them. For example, let us say that you have been on bad terms with your siblings. By understanding the nature of their signs, you can find out how to approach them. Even if they are incompatible with your sign, there are things that you can do to approach them and mend your relationship with them.

That is the power of compatibility. It is not only about giving you options to decide whether you want people in your life or not. It also gives you the knowledge to strengthen bonds with others.

With that in mind, let us dive into the world of zodiac signs.

Aries

Correspondence

Feature	Correspondence
Symbol	The Ram
Element	Fire
Ruling Planet	Mars
Gender	Masculine
Day of the Week	Tuesday
Number	9
Color	Red
Gem	Ruby
Metal	Steel

Body Parts	Head and face
Greek Deity	Athena
Egyptian Deity	Isis
Hindu Deity	Shiva
Roman Deity	Minerva
Tarot Representation	IV - The Emperor

Compatibility

i)   **Aquarius**

They have charming eccentricity that will attract you. In the case of Aquarius, they are characterized by easy-breezy nature and are willing to engage in adventures. As such, it is right to state that there are usually no dull moments when you are with them. They are dependable and reliable. The two of you will be insanely adventurous and wish to engage in anything new that brings you joy. They love doing things as a team and will support you in accomplishing your independent goals. While it is common to find couples that get tired of one another, it is not the case with Aquarius because they tend to want to become friends before anything else. Later when it comes to dating, you will enjoy each other's company and grow strong together. An empowering attribute among them is that they appreciate the solo time and spending time with friends. Thus, the two create a perfect couple.

ii)   Leos

As an Aries, you are also compatible with Leos because both of you share fire signs; you also love games and fun. Leos love keeping you interested by being playful, teasing and flirtatious. Similar to you, a Leo likes to be recognized or validated for what they do and who they are. As a result, both of you will shine brighter together. The relationship will be boosted by each other's confidence and ability to shower each other with affection.

iii)   Geminis

A distinguishing attribute of Geminis, as compared to other signs, is that they are consistently refreshing the interests they have. They are able to bring a new topic whenever you are having a conversation. They are very creative in different ways, and they tend to have great ideas for spontaneous date nights. It is difficult to be bored with a Gemini. Geminis will be attracted to your great sense of autonomy, which means that they won't be clingy but are still reliable.

Traits

- Aries represents the abundance of youthful energy and zest. People are often eager to look for the next adventure or challenge.

- There is a fire within Ariens. They do not back down from competition. And when they compete, they aim to win. They do not tolerate losses easily.

- Ariens love to seek out challenges and dangerous endeavors. This is a way for them to prove their capabilities and show that they are better than most of the people they know.

- You do not need any outside source of energy. You have it within you. But this can often make you act without thinking.

- You are capable of remaining clear-headed in stressful situations. You are quick on your feet and can think fast when the time comes or when it is an emergency.

- Although teamwork does not bother you, it is much preferable to you if you work alone. You do not like being held down by others.

- Often, Ariens can be quite naive and childlike. This can be good, where you show genuine emotions and interest. Or it could be negative because it means you misunderstand or misread situations.

- Ariens know what they want. And they want it now. Actually, scratch that. They should have had it yesterday.

- You do not wait for situations to come to you. You grab life by the horns and deal with situations directly.

Negative Traits

- Just like how the rams 'lock horns,' Ariens can often enter into conflict with others over trivial matters.

- They can be stubborn and do not concede a point easily, even when they are presented with all the facts.

- They do not care what is in their way, as long as the job gets done. This could make them look like people who are not considerate of anyone or anything.

Taurus

Correspondence

Feature	Correspondence
Symbol	The Bull
Element	Earth
Ruling Planet	Venus
Gender	Feminine
Day of the Week	Friday
Number	6
Color	Blue
Gem	Sapphire
Metal	Copper
Body Parts	Neck and throat

Greek Deity	Here
Egyptian Deity	Osiris
Hindu Deity	Shiva (as the sacred bull)
Roman Deity	Venus
Tarot Representation	V - The Hierophant

Compatibility

i) Cancers

The core desires among Cancers are luxury and comfort aspects that you, as Taurus, envy. They are capable of nurturing others and this comes to them naturally. You will love this and end up having a strong connection with Cancer. They are greatly emotional, which in certain cases may turn you off. The good thing about this that they can greatly help you open up your sensitive self and be able to express your feelings and emotions. A Cancer will teach you incredible things about love ranging from buying gifts, dinner and romantic intimacy.

ii) Capricorn

It is a team that forms the most suitable team of roommates and friends. The two are earth signs, and more important is that they appreciate the finer things in life. They are reliable and practical which means that it is difficult for them to fight over chores. The other common aspect between them is their appreciation for organized and clean spaces. Therefore, they will be able to appreciate each other's effort to beautifully decorate or clean their houses, making them compatible.

iii)   Scorpio

A Scorpio is an opposite sign to you, but interestingly there are aspects that make both of you good partners. You will come to value their preference for staying grounded because you enjoy the same. To Scorpios, relationships are investments and it is the same perception for a Taurus. It is the reason you will find that both of you are putting too much energy and time in the relationship. The Scorpios have a no-nonsense attitude which means that when they take on a relationship with you, they will be serious about it.

Traits

- Taureans are all about the simple pleasures. They enjoy their days spending time exploring their senses with music, food, and arts.

- One of your strongest traits is that you know that patience is a virtue. This is why you do not mind waiting for the good things to come your way.

- When you have decided to follow a course of action, there is very little that can steer you away from achieving it. You become committed and focused on reaching your goal.

- There is a sense of calm about you. This is also apparent from the fact that you do not get angry too easily.

- As you are ruled by Venus, you are highly responsive to sexual experiences. This means that you are one who enjoys these experiences completely.

- While change is inevitable, you value stability. You try to avoid falling into too many ebbs and flows or following many changes.

- You value monetary success. You try to keep your endeavors focused on the eventual financial gain that they can give you. You have a flair for making sound financial decisions and developing profitable business ventures.

- When you find something that is valuable to you, then you become possessive of it, whether it is a person, situations, or objects.

- Nature means something special to you. When you are in the midst of nature, you feel at ease and a sense of calm spreads through you.

Negative Traits

- Just like Arians, you can be stubborn when it comes to certain aspects of your life. It is difficult to change your mind when you have made it.

- You have a tendency to be rather covetous. When someone has something that you desire, then you might feel like taking it away from them.

- There are times when you become too self-indulgent, often forgetting about others near you.

Gemini

Correspondence

Feature	Correspondence
Symbol	The Twins
Element	Air
Ruling Planet	Mercury
Gender	Masculine
Day of the Week	Wednesday
Number	5
Color	Silver Gray
Gem	Emerald
Metal	Mercury
Body Parts	Shoulders, lungs, and arms

Greek Deity	Apollo
Egyptian Deity	Twin Merit
Hindu Deity	Various twin deities. For example, the Ashvins, sons of Surya
Roman Deity	Castor and Pollux
Tarot Representation	VI - The Lovers

Compatibility

i)   Geminis

The Geminis are compatible with each other. The reason for this is because they share the same personalities. They are able to communicate in various ways. In the same manner, you are turned-on by mental stimulation. Thus, the two of you enjoy constant communication on different topics. The good thing about being with a Gemini is it brings mutual understanding that one may not achieve with other people.

ii)   Libra

It is right that Gemini and Libra make a perfect couple because they share complex energy. The two have a great and credible desire for life balance, which is required in making a great pairing. The other suitable quality among them is that they are highly social signs. As such, it is easy for them to communicate, chat, and connect, aspects that make a relationship successful. It is a couple that finds living together as satisfying and stimulating to both parties. However, a couple cannot be 100 percent perfect. There are instances where a Gemini has a limit when it comes to personal space and social activities. If you are a Libra having identified this quality, then you will be able to back off and let them have their space when they need it.

iii)   Aries

Aries is full of creativity, a quality that will make you as a Gemini attracted to them. The Aries is full of new hobbies and ideas. There are no good partners to you than those who are trendsetters and trailblazers which means that they can show numerous new experiences making them irresistible. The common attribute between you two is that you are always ready to have a good time and enjoy new experiences. With Aries, you need not worry about getting bored, which may happen with other partners.

Traits

- With the image of the twins, Geminis have a need to have social interactions and conversations. Communication plays an important role in their everyday activities.

- As a Gemini, you enjoy making connections with people, even though you do not like staying too long in one place.

- Your wit and humor are traits that people talk about! You can lighten up the mood in an instant.

- A Gemini is known to be flirtatious, sometimes only for the thrill of it. They know how to work their charms on people.

- Because of the dual nature of their personality, Geminis have difficulty picking sides, even when they feel sure that they have the right choice in mind.

- You are a social butterfly and do not mind surrounding yourself with as many people as possible.

- There is a sense of curiosity within you. This allows you to seek out knowledge about different subjects.

- You might often find it a challenge to stick to any particular thing, whether it is a job, opinion, idea, house, or any other factor. You love diversity and the excitement that new things bring to you.

- The dual aspects of Geminis are present in their personality as well. They are quick to smile and quick to go into bouts of moodiness. This makes their nature rather unpredictable and often, people find it difficult to catch up to them.

- Geminis feel bored easily. This is perhaps due to their nature to seek out new things constantly. But this sense of boredom might affect their mental states as well.

Negative Traits

- Geminis don't like to deal with their emotional fluctuations willingly. This causes their emotions to get out of control.

- They sometimes refuse to look at the consequences of their actions. The idea of ignorance is bliss holds true for the way Geminis approach situations.

- Because they have difficulty picking sides, they also have the tendency to play one side against the other, worsening the situation.

Cancer

Correspondence

Feature	Correspondence
Symbol	The Crab
Element	Water
Ruling Planet	Moon
Gender	Feminine
Day of the Week	Monday
Number	7
Color	Amber
Gem	Amber

Metal	Silver
Body Parts	Stomach and breast
Greek Deity	Apollos
Egyptian Deity	Hormakhu
Hindu Deity	None
Roman Deity	Mercury
Tarot Representation	VII - The Chariot

Compatibility

i) Taurus

Taurus is related to luxury and comfort; combined with your nurturing vibes, it makes a suitable relationship. You should give your caring personality to a Taurus; they deserve it because they are loyal to their partners. You will end up with a stable relationship but you have to be patient because they are slow moving. Although they are not quick in showing it, Taurus has a hidden romantic side, making them suitable partners.

ii) Scorpios

The two signs may appear odd but they are highly compatible. The reason for this is because they share great power in communication and emotional connection. They are different creatures, but both have great strength that enables them to go through various challenges in life. The two share a strong sense of themselves, an aspect that enables them to establish a strong connection that can last for a long time. If you are Cancer, you have compatible personal traits that will enable to connect well with those of Scorpio. As a Cancer, you struggle with certain difficulties while the Scorpio is strong, which ends up creating a perfect balance.

iii)   Virgos

They value perfection in all aspects of life including relationships. The good thing with Virgos is that they are able to conform and adapt to your needs. They are hard workers and will do all it takes to achieve a healthy and happy relationship with you. They will be contented by your emotional support and show of gratitude. It is easy for a Cancer to get in touch with a Virgo's feelings.

Traits

- Cancerians love to forge bonds. They vest their emotions heavily on possessions, people, and places.

- Because of their emotional nature, they find it difficult to let go of objects, people, and places.

- Cancerians love to place their anchors in a place that feels familiar to them. This means that they like to establish roots with family, friends, or within a place that feels like home.

- You might not enjoy taking too many risks. This is because you want to feel safe and secure in your current position. That safety makes you unwilling to change.

- You are tenacious. You don't give up on your goal easily.

- Cancerians are instinctual. When they sense danger or a threat, they withdraw or retreat into their shell.

- Cancerians can read the atmosphere. They can easily pick up on the emotions of others. They are also highly empathetic.

- Because Cancerians are emotional, they can lash out with their emotions. But they only do this when under immense pressure or stress.

- The emotional levels of Cancerians are not always stable. They have their ups and downs. Sometimes, they are able to manage them, while at other times, they cannot do so.

Negative Traits

- Cancerians tend to give more priority to their emotional side than their rational side.

- They often refuse to face the situations presented to them. This is true even when they are about to face conflicts.

- Because they do not like unfamiliar situations, they might not welcome change into their lives easily.

Leo

Correspondence

Feature	Correspondence
Symbol	The Lion
Element	Fire
Ruling Planet	Sun
Gender	Masculine
Day of the Week	Sunday
Number	4
Color	Orange
Gem	Diamond
Metal	Magnesium
Body Parts	Heart
Greek Deity	Demeter

Egyptian Deity	Horus
Hindu Deity	Vishnu
Roman Deity	Venus
Tarot Representation	IV - The Emperor

Compatibility

i)   Libras

As a Leo, you enjoy life and want to be in the company of a person who does the same. You want a person who is passionate about your desires and wants. A good life partner is the one who helps and encourages you to achieve your life goals. Libra makes a perfect partner for you. They desire the same and have a great energy and desire to be social. Both of you will appreciate hosting all of the house parties. When the two of you come together, you always share lively moments.

ii)   Aries

The fact that you both are full of motivation, inspiration, and creativity will help create a strong bond between you. They are able to identify with your intrinsic passion and spirit for life. The two of you will engage in explosive activities and exciting events. You will realize that being in love with an Aries is an unforgettable experience.

iii)   Virgos

The benefit of this sign is that they provide a healthy effect on your life. They are willing and able to assist in various things that you want to be done in your life. There is no better partner in helping you to take care of yourself than a Virgo. What they will need from you is your ability to provide them with gratitude and affection constantly. Your ability to be constant will eventually teach a Virgo to loosen up for love.

Traits

- Those born under this star are playful in nature. They are also good with children and anything that is innocent (like pets).

- Leos are bold. They do not hold back when they want to express themselves. This also means that they are self-reliant and do not like asking for help.

- They are not afraid of being in the limelight. They seek the respect, admiration, and attention of the audience.

- You love to live a life of grandeur, and you tend to project yourself with an air of authority.

- Leos do not like to get their hands dirty. They would rather have someone else take care of the work for them.

- Leos also harbor a deep sense of integrity. To them, aspects of trust and honor are valuable. They not only seek those values but also display them in full.

- You might also like to engage with activities that bring out your creativity. You love to express your talents in unique ways.

- You love to feel special and look for attention from people close to you.

- Leos are confident about their abilities. They are also forward-thinkers and are always looking toward the future and the opportunities that it holds.

- Leos try and find the time for others and usually have the energy to motivate others as well.

Negative Traits

- Leos are egoistic. When they are focused on themselves and their wants, they do not spare time for anyone else.

- There is a sense of arrogance in Leos. They may also come forth as self-centered.

- Because of their strong ego, Leos frequently suffer hurt pride that makes them disappointed in other people.

Virgo

Correspondence

Feature	Correspondence
Symbol	The Maiden

Element	Earth
Ruling Planet	Mercury
Gender	Feminine
Day of the Week	Wednesday
Number	5
Color	Dark blue
Gem	Jade
Metal	Quicksilver
Body Parts	Bowels and solar plexus
Greek Deity	Attis
Egyptian Deity	Heru-Pa-Kraath
Hindu Deity	The Gopi Girls
Roman Deity	Adonis
Tarot Representation	IX - The Hermit

Compatibility

i) Taurus

You share harmonious energy with Taurus. They are hopelessly romantic and hide their sensitive side, and if they get to open up to you then it will bring unambiguous and warm feelings in the relationship. You can count on a Taurus to be reliable. The commitment will be mutual when you are with Taurus.

ii) Capricorn

You two are practical and easygoing in your daily activities. A combination of you and Capricorn results in a calm and collected relationship. The other favorable quality is a Capricorn makes a partner who is sincere and honest. There is no greater couple than the one that shares these noted qualities. As a Virgo, you like to live in a well-kept and organized space. A Capricorn who shares the same qualities will find it suitable to be with a Virgo. Furthermore, Capricorns have thick skin and are confident, which makes easy for them to deal with a Virgo.

iii) Cancers

They value service to other people in the same way you do. Cancers tend to get more personal with things while you prefer being practical. The best thing with them being a partner is that they can help you relax through their unending affection. It is a sort of affection that you may not find from other partners because they make you feel noticed and recognized for your hard work.

Traits

- Virgos understand the relation between the mind and the body. They ensure that they adopt a healthy lifestyle to keep a balanced mind and body.

- They can be innovative with the work or projects that they handle. They can look for ways to approach situations from a unique perspective.

- A system appeals to Virgo more than a sense of randomness. They love to keep order in their homes and in other things in their lives. They do not like the presence of chaos.

- Virgos use their dexterity, skills, and talents to produce results that are nothing short of perfect.

- They can look at the finer details. They also love to be analytical.

- When they receive information, they can easily organize it and use it in the future. They do not get easily bogged by the influx of information.

- Virgos are humble. If you were to invite them to a party, then you can probably expect them to stay back and help you clean afterward.

- Modesty is in their blood. While Virgos can be truly productive, they often like to downplay their accomplishments and skills.

Negative Traits

● Sometimes, they can become so obsessed with the finer details that they forget to look at the bigger picture.

● When they want order, they can become very demanding and controlling in order to get it.

● Despite the fact that they can easily organize information, Virgos can easily become stressed when they are unable to process any more information.

Libra

Correspondence

Feature	Correspondence
Symbol	The Scale
Element	Air
Ruling Planet	Venus
Gender	Masculine

Day of the Week	Friday
Number	6
Color	Turquoise
Gem	Opal
Metal	Copper
Body Parts	Kidneys and loins
Greek Deity	Minos
Egyptian Deity	Maat
Hindu Deity	Yama
Roman Deity	Vulcan
Tarot Representation	VII - The Justice

Compatibility

i)   **Leos**

Leos thrive on the great need for attention, and as a Libra, you are capable of providing it. In return, you will get unconditional love from them. It is right to say that you share a romanticized imaginary perception about relationships. Leo is the partner that you will hold hands with when walking around. They are the kind of lovers who post about you on social media.

ii)   Aquarius

This is a good partner for you. A common attribute between the two of you is that you are of high integrity. However, an Aquarius prefers that you become friends fast before engaging in a romantic relationship. If they let you move into a romantic relationship, they will properly reward your efforts and sacrifices in the relationship.

Traits

- The scales are a representation of balance, and that is what you aim for in your life. You like a sense of equilibrium.

- Librans believe in equality and fairness. They aim to treat others the same way they expect others to treat them.

- Rocking the boat and making people upset are things you dislike doing. If anything, you have charm and grace that you use to your advantage. You like to resolve situations without causing disruptions or conflicts.

- You do not like it when people step out of bounds. To you, the concept of 'fair play' is important.

- When asked to look at a situation, Librans like to look at all possibilities before arriving at a conclusion.

- Loneliness tends to drain the energy out of Librans. This is why they crave companionship and strong bonds.

- You are artistic! This could mean that you could be a painter, musician, instrumentalist, or enjoy any of the many artistic endeavors. You also enjoy excelling in these artistic endeavors.

- During times of conflict, you are willing to let go of your ego in order to make peace with people.

- Outward appearances are important to you, and you will strive to maintain your beauty. To you, the beauty outside reflects the beauty inside.

Negative Traits

- One of the most obvious negative traits of Librans is that they tend to be insincere with other people.

- Librans also tend to make things look good when they are not. This means that they can pretend something has value when there clearly is nothing noteworthy to talk about. For this reason, they can make good salespeople!

- When they are in the presence of authority, they tend to appease them more than necessary to get what they want.

Scorpio

Correspondence

Feature	Correspondence
Symbol	The Scorpion
Element	Water
Ruling Planet	Pluto
Gender	Feminine
Day of the Week	Tuesday
Number	9
Color	Green
Gem	Topaz
Metal	Iron
Body Parts	Genitals
Greek Deity	Ares

Egyptian Deity	Hemmemit
Hindu Deity	Kundalini
Roman Deity	Mars
Tarot Representation	XII - The Death

Compatibility

i)   Taurus

Although an opposite sign of you, the Taurus signifies the same values as you. They too do not like playing games and being flirtatious. They comprehend desire and sensuality in the same manner as you do. The only drawback that the two of you may encounter is that both of you are slow-moving.   When you finally make it to a serious relationship, however, it will be an unbreakable connection.

ii)   Cancers

These are people who comprehend your vulnerability and softness compared to other signs. The two of you have a hard shell which makes it hard for you to open up. It is a common attribute for you to be low key in public, but when in private you enjoy real intimacy. Your love will be 100 percent and they will be willing to go the extra mile to make the relationship work.

Traits

- Scorpios show great strength and willpower toward achieving their goals and beliefs.

- There is an emotional strength to Scorpios. They rely on their will and wits alone to reach their destination, whether in a particular task or their life's journey.

- The idea of showing that they are powerless does not go well with Scorpios. They would rather project a strong facade, despite the situation.

- The phrase 'strong and silent type' would definitely be an apt description of a Scorpio's personality. They do not like to hog the limelight. But that does not mean you take their silence for weakness.

- Scorpios have strong intuitions. They would actually make incredible detectives!

- Researching and digging deep to find out information excite Scorpios. Whether it is in the field of archeology or neurology, Scorpios love to delve completely into their field of focus.

- Because Scorpios have extreme loyalty, they can also be possessive. This could make them jealous easily.

- Scorpios do not accept defeat easily and are prepared to fight to the death (metaphorically speaking, of course).

- You are capable of easily evaluating the emotions of others. This allows you to approach them better.

- Despite loses, Scorpios are capable of easily getting back on their feet and brushing the dust of their defeat. They get back into the fight with renewed interest and energy.

Negative Traits

- Because of their emotional loyalty, Scorpios can harbor deep resentment toward other people. They do not easily forgive.

- When Scorpios feel betrayed, they often seek revenge against the other person or people.

- During certain circumstances, Scorpios can be very controlling and dominating.

Sagittarius

Correspondence

Feature	Correspondence
Symbol	The Archer
Element	Fire
Ruling Planet	Jupiter

Gender	Masculine
Day of the Week	Tuesday
Number	3
Color	Purple
Gem	Jacinth
Metal	Tin
Body Parts	Things
Greek Deity	Artemis
Egyptian Deity	Nephthys
Hindu Deity	Vishnu
Roman Deity	Diana
Tarot Representation	XIV - The Temperance

Compatibility

i)   Aries

Both of you are fire signs which imply the existence of hot passion between the two of you. This aspect makes them a desirable couple because they have an incredible amount of energy that enables them to create a strong relationship. As a Sagittarius and your partner Aries, your relationship will continue to grow strong with time. Your partner appreciates your wild enthusiasm because they possess the same. You are happy when living on your own, and they are independent. The Aries desire adventure which is why they are a suitable fit for Sagittarius. The Aries is there to handle whatever the Sagittarius is feeling, and the fact that they are always busy means that they are suitable partners.

ii)   Aquarius

As a Sagittarius, your perspective and opinions depended on your experiences, and it is hard for any person to tell you otherwise. An Aquarius is a good partner because they are good at detaching. An Aquarius is compatible with many signs including the Sags. The reason for this is that they are happy when hanging out, they take little space, and they tend to go with the flow. Subject to this, they make good romantic partners and living companions. An Aquarius will be happy to let the Sag be himself or herself which is suitable for the relationship.

iii)   Pisces

These are flexible, adaptable, and always ready. At the beginning of the relationship, it may appear like they do not march to your natural fire, but you will eventually get along well. The good thing is that the need for independence is common among you, and because there is an understanding between you, it will strengthen your relationship. Having each other's back will be suitable and an undoubting strength for your relationship.

Traits

- Sagittarians look for the possibility in things. They like to know if there is something more out there.

- They enjoy nature. They feel at peace when they are surrounded by nature. They also enjoy philosophical topics and contemplate about those topics heavily.

- Because of contemplative nature, they love to broaden their minds about topics related to religion, the world, politics, and social science.

- Their love for nature is expressed in their desire to enjoy the great outdoors. They also love to participate in outdoor exercises and activities.

- Sagittarians love to think about the big ideas. They want to understand their life's purpose and meaning.

- Honesty is part of their character. In fact, they can be so honest that often they might just be blunt.

- They are not satisfied with just having enough. They crave more and go out looking for new opportunities to expand, acquire more, or build upon what they have.

- They are typically optimistic, choosing to believe that despite all the challenges in life, things will turn out great in the end.

Negative Traits

- While the presence of optimism brings sunshine and rainbows into the lives of Sagittarians, the absence of meaning in their lives can cause deep depression.

- Because they love freedom, they are afraid of commitment and often find excuses to avoid committing to anything.

- They are also not known for being precise or careful. Often, their carelessness exacerbates problems further.

Capricorn

Correspondence

Feature	Correspondence
Symbol	The Goat
Element	Earth
Ruling Planet	Saturn
Gender	Feminine

Day of the Week	Saturday
Number	8
Color	Brown
Gem	White Onyx
Metal	Lead
Body Parts	Knees
Greek Deity	Pan
Egyptian Deity	Set
Hindu Deity	Yoni
Roman Deity	Vesta
Tarot Representation	XV - The Devil

Compatibility

i)   **Taurus**

A distinguishing aspect with Taurus is that you share many values.

You and Capricorn are both business-minded types and will not pursue something for the sake of it. As such, you will take each other seriously and hold each other accountable. This is because both of you are incredibly loyal and reliable. Taurus is related to tremendous physical pleasure.

ii) Virgos

They are interested in an efficient relationship in the same manner as you. A Virgo is at a high position in meeting your expectations. It is quality that will drive you to be attracted to a Virgo. In their case, they find your hardworking sensibilities attractive. A Virgo will not give up but will look at the soft spots that they can use to get into your heart.

Traits

- Endurance persists among the Capricorns. They do not like to show their struggles. Rather, they quietly endure and aim to reach their goals.

- They can be conservative. They do not like to break the rules. In fact, you could say that they might have created the rules themselves!

- When Capricorns achieve something, they like to see their achievements represented in something tangible.

- You love to prove that you have earned the status and position that you have achieved. This is why you are not afraid of hard work and the challenges it brings.

- Capricorns do not avert responsibility. For this reason, they can be reliable and dependable.

- They understand the value of maintaining their finances properly. They do not enjoy splurging on things until they really have the capacity to do so.

- Whether the path is open to them or not, they make their own way. They are self-reliant and do not ask for the help of others.

- Spontaneity is not appreciated. Capricorns love to plan things ahead and try and take as many factors into consideration as possible.

- They love to have a purpose in life. They do not like to stay wandering from one day to the next without any goal to accomplish.

Negative Traits

- Because of their traits, Capricorns can be really serious at times. They forget to let loose or simply have a good time with others.

- They are strategic because they love to plan things ahead, but this can also make them calculating and ruthless.

- Without a purpose, Capricorns can easily fall into a state of depression or bad moods.

Aquarius

Correspondence

Feature	Correspondence
Symbol	The Water-bearer
Element	Air
Ruling Planet	Uranus
Gender	Masculine
Day of the Week	Sunday
Number	9
Color	Violet
Gem	Sapphire
Metal	Uranium
Body Parts	Ankles

Greek Deity	Ganymede
Egyptian Deity	Nuit
Hindu Deity	None
Roman Deity	Juno
Tarot Representation	XVII - The Star

Compatibility

i)   Aries

They are capable of bringing excitement and fun into your life. The Aries is capable of seeing your flaws as opportunities for a stronger connection. They will be comfortable with your uniqueness and celebrate you. Aries is compatible with you because they, too, are independent and are not easily swayed by others. This common aspect between you will allow you to thrive.

ii)   Libra

You and Libra are air signs and are more inclined to intellectual aspects. As the sign mostly related to brilliance, a flexible and chatty Libra will properly fit with you mentally because they actually understand you. A Libra can be a best friend, a great partner, and an inspiration to try romantic relationships. With a Libra, you can engage in talks for hours which will then help explore a romantic relationship.

Traits

- Aquarians love to become part of communities or groups that have a common goal, interest, or ideal.

- They are focused on trying to bring out something new in the things that they do. Whether it is in their personal or professional life, they want to make a big splash!

- When part of the group, they are the ones who are trying to maintain the cohesiveness of the group.

- Emotions are placed on the sidelines when Aquarians have conversations with other people. They prefer a rational discourse over an emotional one.

- Subjects related to science and technology appeal to Aquarians. They are naturally inquisitive and seek rational explanations for the things around them.

- Friendship is important to them. In fact, they love to have their family and lovers as their friends as well.

- Aquarians understand the rules. They also know the value of the rules. But that does not mean that they might play by them.

- They like to be original when they are expressing their creativity. They want to make their own unique mark on the world.

Negative Traits

- While they enjoy intellectual pursuits, they often have a sense of superiority over others who do not share their opinions.

- Their opinions can be biased because they feel that their knowledge is true.

- They might get demanding at times, especially when they want others to respect their space.

Pisces

Correspondence

Feature	Correspondence
Symbol	The Fish
Element	Water
Ruling Planet	Neptune
Gender	Feminine
Day of the Week	Monday
Number	7

Color	White
Gem	Pearl
Metal	Tin
Body Parts	Feet
Greek Deity	Poseidon
Egyptian Deity	Anubis
Hindu Deity	Vishnu
Roman Deity	Neptune
Tarot Representation	XVII - The Moon

Compatibility

i) Cancer

It is true that Pisces engages in all areas of their living with their hearts. Cancer is someone who will be able to reciprocate your kindness if you are a Pisces. There are times when Pisces become resentful and disillusioned, which is not a good living situation. A Cancer comes in as a person who will give their heart and provide warmth, making the relationship great. The strong sense of themselves that they share makes it easy for them to form a solid and lasting relationship. There is a deep mental connection between Cancer and Pisces which makes the compatibility lasting.

ii) Sagittarius

Sagittarius is intuitive as you. Looking at the qualities between the two of you, it is right to say that you can get into each other's minds and think of something in the same manner. The implication in this is that they will be able to assess how, if they do something, it is likely to offend the other person. In response, they do it in the most suitable way, and the other party will interpret it well. The other suitable quality about them is that they wish to learn what makes the other person tick. They also have the hunger to understand the other soul and body. The quality of respect is essential, their passion credible and they are not afraid of being romantic regardless.

Traits

- The world of fantasy, mythology, and fantastical works of the imagination appeals to Pisceans. This is why they make great storytellers.

- Sometimes, they long to escape from the ordinary and mundane and become part of something that they consider unique.

- They love to be the saviors. Often, they think of ways to contribute to the world at large.

- There is a constant need to feed their creative talents. Otherwise, it is as though they are like a fish out of water where they feel a sense of suffocation.

- They tend to be the more romantic one in a relationship. They work hard to retain the magic.

- Pisceans try and contribute toward charity and participate in non-profit organizations. They are selfless in their endeavor to help others.

- They understand the emotions of other people. This is why they approach a situation with tact and care.

Negative Traits

- They might be rather elusive. This means that they do not easily commit to a long-term relationship.

- Because of their nature, they can be taken advantage of easily by others.

- While they can be independent, they are often seeking to be 'rescued' by others, making them rather dependent.

Conclusion

Life is filled with mysteries.

But sometimes you receive a codebook that can help you solve some of those mysteries. Sometimes, you might be able to gain great insights and life lessons. Sometimes, you might reveal something transformative about yourself. Other times, it might be about the people, situations, ideas, positions, status, health, and other factors that are part of your life.

Numerology and astrology are your Wikipedia, library, and sources. The only difference is that they are focused on your life, your goals, your dreams, ambitions, and everything else that is part of your life.

It is the source of knowledge that guides you.

And as they say: Knowledge is power.

What you do with this knowledge is entirely up to you.

Become what you are destined to be.

Make the change you want to see.

And May the numbers and the stars align your way.

Printed in France by Amazon
Brétigny-sur-Orge, FR